VEGETABLES

VEGETABLES

A COMPLETE GUIDE TO THE CULTIVATION,
USES AND NUTRITIONAL VALUE OF
COMMON AND EXOTIC VEGETABLES

ANN BONAR

GUILD
PUBLISHING
LONDON

This edition published 1986 by Book Club Associates by arrangement
with Nicholas Enterprises Ltd.

Designed and produced by Nicholas Enterprises Ltd
70 Old Compton Street
London W1V 5PA

Editor: Jennifer Mulherin
Art direction: Tom Deas
Special photography: Graham Young
Home economists: Linda Fraser, Jennifer Mortimer

Published 1986 by
Hamlyn Publishing,
A division of the Hamlyn Publishing Group Ltd.,
Bridge House, London Road, Twickenham, Middlesex, England.

Printed in Belgium

Contents

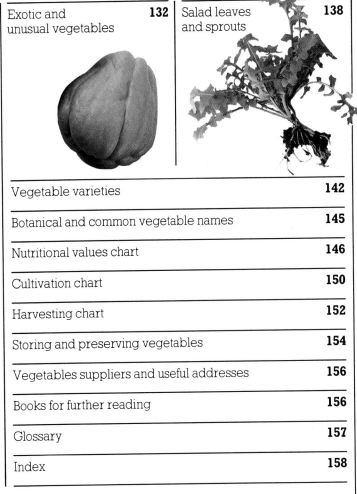

VEGETABLE ENCYCLOPAEDIA
Artichoke to Turnip

All vegetables shown in this section are actual size.

Apium graveolens **70**
Celeriac

Brassica napus **120**
'Napobrassica'
Swede

Brassica oleracea 'Italia' **64**
Calabrese

Apium graveolens 'Dulce' **72**
Celery

Brassica oleracea **60**
Cabbage

Brassica rapa **130**
Turnip

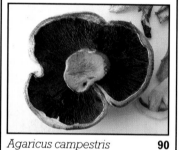

Agaricus campestris **90**
Mushroom

Asparagus officinalis **48**
Asparagus

Brassica oleracea 'Botrytis' **68**
Cauliflower

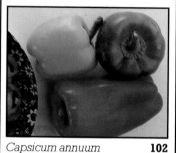

Capsicum annuum **102**
Sweet pepper

Allium ampeloprasum **82**
'Porrum'
Leek

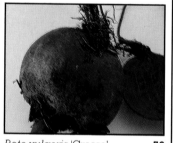

Beta vulgaris 'Crassa' **56**
Beetroot

Brassica oleracea bullata **58**
'Gemmifera'
Brussels sprouts

Chichorium endivia **78**
Endive

Allium cepa **94**
Onion

Beta vulgaris 'Cicla' **124**
Swiss chard

Brassica oleracea 'Cymosa' **118**
Sprouting broccoli

Chichorium intybus **74**
Chicory

Introduction

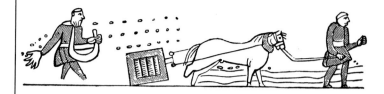

Gardeners tend to be divided into two groups: those who are attracted to ornamental plants, and those who deal with the edible species. The latter includes fruit, vegetables and herbs and here again there are sharp, even passionate, divisions into fruit, vegetable and herb specialists. The largest of these groups at present is the vegetable grower.

Vegetables, along with herbs, were the first plants to be deliberately grown by man in the soil outside his backdoor (or perhaps in the soil in front of his cave). They provided easily-accessible, nourishing food, and some which can still be found growing wild today include carrot, celery, asparagus, cabbage, turnips, leeks and parsnips.

Others that have been cultivated in Europe for many hundreds of years include broad beans, peas and onions. Beans of all kinds have always been a worldwide vegetable; lettuce and radish are European vegetables of long standing. In the sub-tropical areas of the Mediterranean there was and still is much greater variety, and the Romans introduced a good many to Britain 2,000 years ago, amongst them probably globe artichokes, Florence fennel, cucumbers and endive.

Even greater variety is found in tropical regions, and from the Americas such vegetables as tomatoes, potatoes, pumpkins, marrows, sweetcorn, aubergine and peppers have eventually found their way into British and northern European gardens, to be grown by gardeners determined not to be beaten by the weather. Corn or maize was a staple part of the diet of the American Indians, tomatoes grow wild in South America, and potatoes were so much a part of the Aztec everyday life that pottery containers shaped like them have been found in excavations.

Vegetable cultivation in Britain

The Roman invasion of Britain, southern Europe and north Africa brought civilization to these lands. The Romans were great epicureans, enormously interested in food and its flavours, and it is probable that the wide variety of vegetables now grown throughout Europe owe their development to them. In Britain alone, there cannot have been less than twenty-five different kinds of vegetables grown during their occupation, probably more. In the centuries between their departure and the arrival of the Normans, vegetable growing undoubtedly declined with the repeated invasions of Britain from northern Europe. The cultivation of vegetables became confined to the gardens of the religious houses, along with herbs and fruit.

The mediaeval centuries saw a gradual onset of peace and order. Even peasants began to own pieces of land and no longer had to rely on access to common land for grazing and cropping. Word spread from the monasteries and the manor gardens as to the best ways to grow vegetables, and the best kinds would have spread by seeds, and pieces of plant passed from one neighbour to another.

By the time Jon the Gardener wrote his famous gardening poem in the later part of the fourteenth century, he was able to write about onions, leeks, and cabbages (wortys). Even as far back as that, cabbages could be grown all year round, with care and attention to sowing times, listed by him and summed up as:

'And so from month to month
Thou shalt bring thy wortys forth.'

In the record of the accounts for the London garden of the Bishops of Ely for 1372, leeks, onions and 'beans in the husks' are mentioned; for instance: 'And of 16s for onions and garlick sold' as produce from the garden.

The most popular vegetables of those times were the three mentioned above, together with kale, cauliflower (then called colewortes), turnip tops, a type of cabbage that could be cut down to the stump but would still sprout to produce leaves for summer use, and the pulses: peas and beans. For the majority of people the mediaeval diet was very restricted and consisted largely of soups, bread and vegetables, with cheese, fish and meat as rare treats. Flavour was at a premium and the more strong-tasting the vegetable, the

more it was grown and used, hence the considerable popularity of onions, radishes which were the only root crop grown to any degree, and garlic which was eaten in large quantities throughout the country.

Turnips and carrots were grown, but the former mainly for their tops, and the latter were smaller and less well-flavoured; both also went into the stews. Parsnips had a much more pronounced taste even then, and were eaten separately or as a sweet dish, with apples.

While vegetables were a staple part of the diet in early mediaeval days, their use gradually declined towards the end of that time, so that: 'they were supposed as food more meet for hogs and savage beasts to feed upon than mankind.' according to Holinshed, a historian writing during Elizabeth I's reign. But vegetable cultivation revived in the 16th century when the exploration and discovery of new lands, and the New World in particular, brought seeds and plants of all kinds pouring into Europe in an ever-increasing flood. Amongst them were potatoes, tomatoes, aubergines, sweet-corn and peppers, both sweet and hot, and the merchants, gentlemen and nobility: 'made their provision yearly for new seeds out of strange countries.'

Range of vegetables in 16th-century Britain
The range of vegetables became much greater and included cucumbers, radishes, skirrets, parsnips, carrots, cabbages, turnips, onions, leeks and a variety of salads such as lettuce, chicory, endive, sorrel, spinach, rape, orache and rocket. Skirret was widely grown; its botanical name is *Sium sisarum*, and as a native of northern Europe it was constantly mentioned in records of the times. Its roots were the important part, being fleshy tubers with a sweetish taste. Another surprising vegetable that seems to have been commonly grown was the sweet potato, of South American origin. Gerard, the herbalist, grew it in his Lincoln's Inn garden, where it apparently 'flourished unto the first approch of Winter'; he described the roots as: 'many, thicke and knobby . . . joined together at the top into one head, in maner of the Skyrret.'

Ordinary potatoes, then known as potatoes of Virginia, were, however, hardly grown although Gerard said they were good and wholesome, and could be roasted or boiled and eaten with oil, vinegar and pepper. Enthusiasm may have been lacking because they were known to be associated with poisonous plants, and it is a fact that any green parts developed by a tuber are toxic, and should never be eaten. Both sweet and hot peppers rapidly became popular, especially the hot ones whose strong flavour was relished by the Elizabethans.

From this time onwards, vegetable cultivation continued and increased; it became more and more sophisticated, and recipe books detailing methods of cooking vegetables began to appear. Jerusalem artichokes, called potatoes of Canada, became so widespread that they were eventually thought only fit to feed to poultry and swine. Peas were sown in summer for autumn cropping, and in early autumn for picking the following spring; pea varieties included the sugar pease, the gray, the spotted, the pease without skins, and even chick peas. There were eight different sorts of kidney beans.

Recipes for cooking and using vegetables
In a book written in the seventeenth century, it was said that potatoes (presumably the sweet kind) would make very good bread, cakes, paste (pastry) and pies; they 'increase of

Above Cultivation of sweet potatoes in Kentucky, US c. 1885.
Left London street seller with cucumbers c. 1820.
Below Costermonger selling cabbages c. 1888.

Seven varieties of aubergine, illustrated in a gardeners' handbook of 1900.

themselves in a very plentiful manner with very little labour. Vegetable pies and tarts became fashionable; for instance spinach tart flavoured with cinnamon and rose water, or made with sorrel, parsley and eggs. Lists were made of salad vegetables and included onions, lettuce, samphire (a fleshy coastal wild plant), asparagus, cucumbers, radishes, skirrets and carrots, as well as red sage, mints, lettuce, violets and marigolds. A special salad recipe for a feast consisted of 'cabbage done with cucumber, currants, orange, lemons, olives, figs and almonds.' This was not an unusual mixture in those days when savoury and sweet flavours were frequently combined.

Protected cultivation

During the eighteenth century the outstanding development in vegetable gardening was the arrival of protected cultivation. The introduced vegetables of Elizabethan days were often hot-country plants and needed so much care in cultivation that they were restricted to the expert private gardener of some standing. Such vegetables were not grown on a commercial scale, but with the new methods of manufacturing glass that appeared at this time, glass structures and frames could be built, to which the heating used for orange trees was added. Vegetable seasons became greatly extended. Cauliflowers were grown under bell-glasses, the forerunner of the modern cloche; asparagus was grown beneath frames, and in late spring in the market gardens round London, radishes, lettuce, spinach, beans, peas, cabbages and chives were some of the vegetables being cropped.

This type of protected cultivation required shelter from wind and a sunny position. This lead to the development of the square walled garden, the grand kitchen garden of the Victorians. A square walled garden was no new idea; the gardens of the Middle East 4,000 years ago were walled, and although walls were replaced at one time in Britain with hedges, the walled garden came back into favour because, once built, it provided greater shelter and less labour. Within this, the crops grown for the great house were many and varied, with vegetables needing to be in constant and varied supply all year round.

It was at this time that the tomato or love-apple as it was still known, at last came into its own. Gerard had advised that the seed should be sown in a bed of hot horse-dung, and with the advent of glass protection it became easy to grow it to the fruiting stage. Similarly aubergines – the 'madde apples', *Mala insana* – were a curiosity until the plants could have the shelter and thereby the warmth to which they were accustomed in their native habitat, although they were known to be widely eaten in Arabic countries and in southern Europe.

A good deal of forcing and blanching went on, whereby cardoons, seakale, chicory and endive were made to produce their shoots early and without colour for winter eating. Rhubarb, cucumbers and potatoes were forced; mushrooms were available all year, even lentils were grown.

Extending the vegetable range

But what was the most staggering was the number of varieties of each vegetable, as well as the different kinds. A book on vegetables and their cultivation published at the time fills 597 pages. It is profusely illustrated with woodcuts of vegetable varieties, showing all kinds of shapes and sizes to a degree that is unheard of today.

There are 13 pages devoted to cabbages, for example, with 47 varieties illustrated; lettuce extends to 14 pages with illustrations of 48 varieties, with names such as Brown Geneva cabbage lettuce, White Silesian, Curled Californian, the Spotted Cos and the Red Besson. Unfamiliar vegetables include alexanders, *Smyrnium olusatrum*, displaced by celery, the Chinese yam, skirret – the mediaeval vegetable – samphire, rampion *(Campanula ranunculus)*, sweet potatoes, orach for its leaves, and evening primrose for its roots. This book shows that there are enormous untapped possibilities for a far greater range of home-grown vegetables to expand the nutritional value and increase the number of flavours of vegetables. Add to this the foreign vegetables now imported all year round and an extremely varied and imaginative diet based on vegetables can be followed, supplying food which is nutritious and satisfying, but without the health risks associated with animal fat and protein.

Vegetables for health

If it were not for plants, mankind would die. Not only that, everything that lives – mammals, insects, birds, reptiles – ultimately relies on vegetation. Man eats beef, lamb, pork and bird flesh, but what do they all feed on? Grass, other wild plants, grains and seeds.

It is universally agreed that plants in some form must constitute part of our diet. In many English speaking countries, alas, vegetables are regarded simply as fillers for more expensive meat or fish. Certainly they are considered boring to eat, being dull and flavourless because of unimaginative cooking methods. In some European countries, notably the south, the range of vegetables is wider and their preparation interesting and varied. On the whole, however, they are meant to complement meat, fish or game rather than be a substitute for them. Yet in many Eastern and Oriental countries where fresh meat and fish is not so generally available, vegetables can be found to form an intrinsic part of the cuisine. Western nations can learn a lot from such countries and the current interest in Eastern and Oriental food is healthy in more ways than one.

Vegetables are exceedingly nutritious; all can be delicious to eat, and can easily provide a satisfying meal. The harm in eating vegetables has yet to be discovered, whereas meat, dairy products and sugar, which play a large part in Western diets, can cause trouble even in a mixed diet. If we turn our view of eating food inside out and build our menus round vegetables, making them take the place of meat as the centrepiece of a dish, the benefits to our health and vitality would be enormous.

Nutritional value of vegetables

A glance at the nutrition table for vegetables at the end of this book will show that they contain most of the elements needed for human diet, some in small quantities, some in large. Their greatest strengths lie in the mineral and vitamin contents, though many have appreciable percentages of dietary fibre, and practically all have a high water content. This last is more important than it would appear, as few of us drink the amount of water we should every day, and many of the summer salad vegetables are, literally, thirst-quenching at a time when human water loss through perspiration is at its highest.

Minerals

At least 18 minerals are necessary to good health, and vegetables contain most, if not all of these.

To name a few minerals, *calcium* is important for teeth, bones and good blood; highest quantities will be found in spinach, okra and leek. *Iron* is universally known to be essential for red blood corpuscles and is particularly found in leafy vegetables with dark green leaves, such as spinach and endive, and also leek and radish. *Phosphorus* works in association with calcium, and helps in the release and use of energy; without it some of the B-vitamins cannot work, and it is a constituent of nucleic acid. It occurs most in the pulses. *Magnesium* is found in bones, and is an essential part of every human cell. Because it is a part of chlorophyll, the green colouring matter of plants, it will be absorbed automatically every time the leafy or stem vegetables are eaten, and also some of the fruiting vegetables, such as marrows, sweet peppers, peas, beans and cucumbers.

Potassium is contained in the body fluids and works with *sodium* to ensure efficient cell functioning; sodium is also concerned with the body's water balance and is essential to the working of muscles and nerves. In Western diets, a great deal of sodium is absorbed as salt; this results in a low potassium intake. Salt is associated with the onset of high blood pressure. Vegetables, however, have a high potassium: low sodium ratio. In other words, the reverse of what is normally eaten is the state in vegetables. Practically all vegetables contain relatively high quantities of potassium and little sodium. Eating plenty of vegetables regularly should therefore help to restore and maintain the balance of these two minerals within the body.

As with plants, some minerals are necessary to health in

large amounts, some only a little, and others are trace elements, required only in quantities of the order of parts per million rather than grams or milligrams; such elements will be automatically present in wholefoods.

Fibre in vegetables

Fibre or roughage is essential for good health, and it is believed that the absence of fibre in a regular diet can lead to such illnesses as chronic constipation, diverticular disease and bowel cancer. Without it, the human intestine works sluggishly so that waste products accumulate and in time become toxic. A large intake of fibre ensures that the products of digestion pass rapidly through the bowels and that the intestine is constantly exercised.

All vegetables contain some fibre, much of it in the outer covering of skin or rind or in the stalks which unfortunately are often discarded while preparing for cooking, and incidentally along with them some of the vitamin content, particularly vitamin C. More is present in grains such as brown rice, or wholemeal flour, but if vegetables are eaten with their skins or the stalk, then a useful amount of fibre is absorbed. Vegetables like celery, Swiss chard, leeks, sweetcorn and beans naturally have the highest fibre content, which is eaten automatically as it is an integral part of the edible portion. Tomatoes have little fibre, but removal of the skin halves what there is; potatoes baked and eaten with their skins have twice as much as the peeled, boiled version.

Vegetable protein, fats and carbohydrates

Protein, in the form of meat and dairy produce, is a major constituent of the average diet and for many years it was believed that a high protein diet was 'body building'. Protein *is* essential for good health but recent medical research has shown that too much animal protein and fat is linked with heart disease and some forms of cancer. All nutritionists now recommend a decrease in the consumption of animal products and an increase in vegetable protein and fats.

Protein is not high on the list of vital ingredients present in vegetables, with the exception of what are known as the pulses: beans of all kinds, peas and lentils. Lentils have the highest protein content, followed by peas, but haricot and butter beans contain a great deal if eaten raw, and both peas and other sorts of beans are better eaten raw when young, which they easily can be if included in salads. The pulses also contain a good deal of fibre and little fat, so they must rank as No 1 in the vegetable food charts. However, the protein content of other vegetables is not negligible; all contain some, and those with the higher figures are surprising, for instance, spinach, sweetcorn, and Brussels sprouts.

Fats are generally in short supply, so are carbohydrates such as starch and sugars, except in potatoes, parsnips and some other root vegetables. Nutritionists advise vegetarians to use corn and nut or fruit oils and whole grains in the form of bread, pasta and rice as substitutes.

Vegetables all year round

It is not difficult to maintain a steady supply of fresh vegetables all year round from one's own garden, and even with small areas, it is surprising how much can be fitted in with careful preliminary planning and intensive cultivation. Even the city dweller, with only a window-ledge or two or a balcony, can still grow some vegetables in pots and tubs, especially as new varieties have been bred for containers, and the gardener who has a basement area or roof-garden can work miracles with such spaces.

The summer and autumn months are not difficult to fill with such a variety of vegetables that it amounts to a glut; carrots, lettuce, beans of all kinds, spinach and Swiss chard, tomatoes, cucumbers, aubergines, sweet peppers, peas, fennel, courgettes and marrows, the first sprouts and autumn cabbage. Early summer sees the cropping of asparagus, the first lettuce, beetroot and radish, new potatoes, peas, summer cabbage, young turnips and spinach, broad beans and spring onions.

In winter there are the root vegetables, such as parsnips, stored carrots, beetroot and celeriac, potatoes and onions; winter celery comes into its own with the frost, as do Brussels sprouts and the Savoy cabbages. Red and white winter cabbage are also available, so are Jerusalem artichokes, swedes, winter radish and cauliflower. Stored marrows, pumpkins and squashes will keep well for several months.

Leeks will start to be cropped in winter and continue through into spring and kale is another immensely hardy, winter-spring vegetable; in sheltered gardens, sprouting broccoli may be ready as early as late winter and can go on until late spring. Perpetual spinach will start to grow again in spring, and onions will last until spring.

Hungry gap

In northern temperate countries, the most difficult period for fresh vegetables is from March to the end of May, often known as the 'hungry gap', and a list of those available at that time is provided on page 44. The other difficulty is not in numbers but in the type of winter salads available. These have the advantage of supplying vitamins, especially vitamin C, at a time when these are at a premium. Blanched chicory can be obtained from December onwards; blanched endive will continue up to Christmas outdoors or with protection; the hardy chicories will survive if they are the red varieties known as radicchio, or can be planted in a greenhouse that might otherwise stand idle through the cold weather. Winter radish varieties, much larger and hardier than the small red summer kinds, can form part of a winter salad as can the florets of cauliflower, celery and grated carrots, celeriac, Dutch cabbage and red cabbage, and Hamburg parsley. Corn salad (lamb's lettuce), winter cress (like watercress) and chervil are three small leafy herbs which will add piquancy and a refreshing flavour.

Importance of freshness

The quality of freshness in vegetables is most important. They are at their most nutritious if cooked and eaten immediately they have been gathered, and better still if eaten raw, as cooking inevitably destroys or changes some of the vitamins and proteins and results in some of the minerals being leached out altogether. Many vegetables eaten as soon as picked have a quite different flavour to those a few hours old, others have a more intense flavour when fresh. All in all, garden vegetables have a great deal to contribute, not just to a luxurious or gourmet meal, but to the everyday food which keeps us alive.

The treatment of any vegetable to retain the maximum flavour and nutrition from the moment it is picked until it is served is more or less the same. The interval between harvesting and consuming should be as short as possible, and if it is to be cooked, that between harvesting and cooking should be similarly brief – straight from the garden to the pot is ideal in most cases, especially for the leafy vegetables. If they have to wait a few hours, they are best stored wrapped

All vegetables, including winter vegetables such as potatoes (**above**)and leeks (**right**), are at their most nutritious when cooked soon after they are picked. Freshness is especially important in leafy vegetables such as spinach (**below**) which require a minimum of cooking or sometimes none at all.

Above Commercially produced crops, such as celery, are often days old before they reach markets or stores and have consequently lost many of their nutrients.

in clean polythene sheets in the salad compartment of the refrigerator.

Cleaning and preparing vegetables

When cleaning and preparing vegetables, remove as little skin as possible; use stalks cut up finely so that they cook more quickly, and tear leaves into small pieces rather than cut them. Pat them dry if washed before storing, otherwise some of the contents will be lost by seepage into the water. But some vegetables need to be left damp to retain their crispness, for instance celery or runner beans; if lettuce are to be kept, they will remain close to their original condition if harvested complete, with the root still attached, placing this in water or wrapping it in damp polythene. Light destroys vitamins, so keeping vegetables in the dark is another important point.

Storing vegetables

Never store vegetables, even for a few hours, with obvious rots, injuries, bruises or pests on them. Bacterial soft rot, for instance, can destroy a root vegetable such as a carrot within a matter of hours, and spread to those on either side of it. Many root crops have a better flavour and are more nutritious if they can be left in the ground in the winter; paradoxically, new potatoes taste better if allowed to mature for a day or two after lifting; onions which are to be stored have to be ripened of course, but correctly done, this helps to seal in the nutrients. Fruiting vegetables start to change their starch into sugar as soon as they are picked, so the longer they are kept, without maturing to the point of decay, they will be correspondingly sweeter.

Cooking methods

If vegetables are to be eaten raw, the preparation should be speedy, with the minimum of tearing, cutting or grating. For cooking, vegetables can be boiled, but in as little water as possible without actually burning them, preferably without salt – which can be added later if required, according to individual taste. Steaming vegetables over boiling water is a better alternative; not only do they retain more nutrients and flavour but they hold their shape and colour better. Boiled or steamed vegetables should be cooked for the minimum possible time. In general, they should be tender but still firm and a little crunchy. This applies particularly to such green vegetables as cabbage, French beans and courgettes.

Braising is another good cooking method, in which the

vegetable is gently cooked in a little butter or oil to which stock can be added. Baking, roasting and grilling vegetables are also popular cooking methods which seal in goodness and flavour.

If sliced or finely cut, they can be stir-fried by the Chinese method in a little vegetable oil; quickly and easily done, this is one of the healthiest and most delicious ways of cooking vegetables. Vegetables simmered in water over a long period are not very nutritious or flavourful except in soups and casseroles, when the liquid forms part of the dish.

For health, fried vegetables should be avoided; however, if you must, use a single vegetable oil such as corn, sunflower, or soya, not a 'blended' vegetable oil, which contains a substance called a trans fatty acid; this prevents essential fats in the body, sometimes called vitamin F, from metabolising normally.

Using vegetables imaginatively

Some of the great dishes of the world are based on vegetables. Minestrone, for example, is essentially a vegetable soup; ratatouille, one of the most delicious of vegetable stews and pommes Dauphinoise, perhaps the greatest potato dish ever invented. What is notable about the world's renowned cuisines is that they all treat vegetables with respect. Cooked simply or elaborately, vegetable dishes are designed to emphasise or enhance the natural flavour of the vegetable itself. The Chinese, for example, place great store on retaining the texture and colour of their vegetables and use them in subtle combinations with meat and poultry to produce dishes which are both a delight to the eye and the palate.

It is only recently in English-speaking countries that vegetables have been accorded their due respect – partly for health reasons and partly because of an interest in foreign cuisines.

To some extent this neglect has been because vegetables native to cool temperate climates are not large in number, but with the use of protection in the form of cloches, tunnels, frames and greenhouses, a much greater selection is possible. Improvement in cultivation methods and breeding of strains hardier or specifically suited to a particular season has enabled much more round the year growing with the result that many fresh vegetables – well over fifty distinct varieties and types – can now be grown at home.

Vegetables can form the main course of a meal more than adequately, and should do so at least once a week for the average family. Protein will be supplied mainly by the pulses; crisp vegetables like celery, fennel, carrots, sweetcorn, radish and Swiss chard stems can provide texture and crunchiness. Strongly-flavoured vegetables such as onions, tomatoes, parsnips, aubergine and Jerusalem artichoke will ensure tastiness, and any of the green leafy vegetables provide chlorophyll and therefore iron. The addition of an ingredient like cheese, herbs, eggs, mushrooms or nuts can produce satisfying and delicious dishes. Risottos or different types of pasta served with a vegetable sauce – for instance, onion and nutmeg, tomato with garlic, or mushrooms with cream and sherry – will delight the most seasoned gourmet.

There are also many luncheon and supper dishes of vegetables; what could be nicer than an asparagus flan, or an egg ratatouille or piperade? Omelettes lend themselves well to vegetable incorporation, such as tomatoes or onions and there is the Spanish version (tortilla) which usually contains potatoes as one of the ingredients. Pancakes are another mouth-watering possibility, and pastry made with whole-

wheat can form the basis for spinach tartlets, mushroom pastries or vegetable pies.

As has been indicated, vegetables as side-servings to accompany the main dish can be cooked in all sorts of ways which make the most of their flavour and nutrition; they can also be served cold as salads, either raw, or lightly cooked, and there are endless variations on this theme. Again, it is a case of being innovative and imaginative, of trying different combinations of leaves and other salad vegetables. Use them with dressings and sauces which go beyond mayonnaise and vinaigrette; try ravigote, a vinaigrette with chopped mixed herbs added, and gherkins or anchovies; use chopped watercress or spinach and fresh herbs to make sauce verte (green mayonnaise) or make a sweet and sour dressing with ginger, soya sauce and wine vinegar or lemon. Try American dressings such as Blue Cheese or Thousand Islands. Give children celery or carrot to crunch instead of biscuits, and wrap cabbage or lettuce leaves round a stuffing to provide a version of the Greek dolmathes.

Cooking vegetables Steaming (**left**) is probably the healthiest way of cooking vegetables. Boiling (**above**) is a more conventional method but some of the nutrients are lost in the water. Braising (**above right**) and stewing (**right**) are also common cooking methods which retain much of the vegetables' flavour and goodness.

Vegetarian diet

A totally vegetarian diet is no hardship, as should be obvious by now, and for flavour and variation it can be the equal of a conventional meat-containing one. Nutritionally, it is probably superior provided there is sufficient protein in the diet. The daily intake of protein for adult men in a sedentary occupation is said to be round about 60g (2oz), for most women 55–60g (1¾–2oz) and for children between 4 and 11 years old, about 40–50g (1¼–1¾oz); protein contents of individual vegetables can be obtained from the table of nutritional values given on pages 146–149. A well-balanced diet can be had from the four main groups of foods that constitute the average vegetarian diet: fruit and vegetables; dairy produce; grains; and pulses, beans and nuts – the last-named contain most protein. In the light of today's findings that saturated (animal) fats are likely to be a cause of illness, dairy products can either be substantially reduced or substituted by vegetable cheese, skimmed milk, polyunsaturated margarine and vegetable oils.

Protein-rich vegetables

In a vegetarian diet, it is important to make sure that sufficient protein is eaten. It is contained in fish, meat, eggs and dairy products and in whole grains, nuts, seeds, dried legumes (pulses) and some vegetables. Vegetables with the highest protein content are listed below.

grams/100 grams

Asparagus, boiled	3.4	Pea, frozen, boiled	5.4
Beans, broad, boiled	4.1	Pea, tinned	4.6
Beans, butter, boiled	7.1	Pea, dried, boiled	6.9
Beans, haricot, boiled	6.6	Potato, fried	3.8
Beans, runner, raw	2.3	Potato, old, roast	2.8
Broccoli tops, boiled	3.1	Potato, old, baked	2.6
Brussels sprouts, boiled	2.8	Potato, crisps	6.3
Cabbage, savoy, raw	3.3	Spinach, boiled	5.1
Cabbage, winter, raw	2.8	Sweetcorn, boiled	4.1
Mushroom, fried	2.2	Sweetcorn, tinned	2.9
Pea, raw	5.8	Turnip tops, boiled	2.7
Pea, boiled	5.0		

Vegetable variety

In order to be able to eat vegetables which will supply a wide range of nutritional needs and which will be fresh all year round, it is necessary to widen one's sights, to deviate from the well-worn track, and to explore some unfamiliar side-turnings. It means experimenting, and although there may be some failures initially, it will produce some pleasant surprises, as new varieties are found to be, not only unexpectedly flavoursome, but also no trouble to grow.

Of course, there is nothing unacceptable about the commonly grown vegetables; they would not have been cultivated for hundreds of years if they had not been easy to grow and had some food value and flavour. Our ancestors grew cabbages, beetroot, broad beans, leeks, onions and turnips; regrettably many vegetable gardens still only contain these few.

However, there are so many more to try which can brighten up our food, enhance its flavour, and improve its nutritional value, that idleness or lack of space are no excuse for their absence from any self-respecting kitchen garden. For instance, why grow cabbage, which is attacked voraciously by caterpillars every year and is prone to diseases which are resistant to chemical control, when perpetual spinach is immune to practically everything, and provides good greens containing iron all year round? Or Swiss chard, which provides two vegetables for the space taken up by one, for most of the year?

If lack of space is the difficulty, grow small varieties and grow the plants closer together. You will probably not get such gigantic or perfect vegetables, but they will be perfectly acceptable to the average family, and you will have the satisfaction of knowing that the much greater choice provides a more complete cover of minerals, vitamins, fibre and proteins, and in much greater quantity.

To achieve this spread, it is necessary to go beyond the standard list of brassicas – cabbage, cauliflower, Brussels sprouts – and root crops such as carrots, beetroot and radish. Broad and runner beans, and such salads as lettuce, tomato and onions are useful but unimaginative as are new potatoes, marrows and leeks.

Easily-grown winter alternatives

Easily-grown hardy and familiar alternatives, or additions if desired, could include the spinaches, preferably perpetual spinach (spinach beet) and Swiss chard. (The winter and summer forms of spinach itself have a better flavour but their tendency to run to seed is their drawback.) *Swiss chard* has large, shiny, spinach-shaped leaves and crisp, wide, white stalks which can be cooked to provide a separate and deliciously crunchy vegetable. Its red form, Ruby chard, has the same virtues and is decorative into the bargain, with its crimson stems and red-tinted, dark green leaves. *Perpetual spinach* will provide leaves in winter if protected with cloches, not necessary in mild gardens or warm climates. New Zealand spinach is another type, which although not hardy, is extremely useful for a hot corner where the soil is dry, and where it will simply grow and increase without trouble. Unlike the other varieties, it trails along the soil, with much smaller leaves, and is rather fleshy; the growing tips, as well as the leaves, can be used.

More greens for the winter could include both purple and white *sprouting broccoli* and the perpetual kind. Admittedly, they do stand in the ground from spring to mid-winter before they can be used, but in the back-end of winter when vitality is at a low ebb, they will provide fresh 'greens' from March and often earlier, to sometime in May, thus covering a large part of the period when vegetables are at their shortest. Perpetual sprouting is even more useful as there is no need to sow and plant every year; it remains in the same place and with regular spring mulches of organic matter will crop every winter.

Leeks are another invaluable standby for winter greenery; if sown late in spring, they, too, will continue to crop until late in May. *Kale* is a tough and hardy vegetable which will provide green leaves all winter, if the right variety is grown,

and has shoots like sprouting broccoli in spring as well. Even the common cabbage can be varied, if *Savoys* and *red cabbage* are grown, both providing leaf in winter.

For winter root crops, if parsnips are not popular, and their flavour is not to everyone's taste, there are swedes, *Jerusalem artichokes*, which grow themselves, and *celeriac*, the celery-flavoured turnip, similarly easy to grow.

Alternative summer vegetables

Some summer vegetables which are expensive to buy but not to grow include *sweet peppers*, *aubergines*, *Florence fennel*, which even in bad summers will still bulb up if given a sunny site, *calabrese* – the summer form of sprouting broccoli – and peas, including their increasingly popular variety, *mange-tout*.

Other luxury vegetables include *asparagus*, a perennial, long-living plant which needs quite a lot of care but is very well worth it for its flavour, duration of cropping season – six-eight weeks – and cost-saving. *Globe artichokes* are a delicacy, too; they will crop for at least three months and have a life of four-five years. If space is short, they are decorative enough to grow in a herbaceous border.

Salad vegetables

The possibilities for salad vegetables allow no excuse for unimaginative salads either in summer or winter. Outdoor cucumbers are no more difficult to grow than marrows, and will provide fruit from late in midsummer up to mid-autumn, longer in good seasons. *Endive* will supply an interesting variation on lettuce, and the summer form of celery can be added to salads, so can *Chinese cabbage*, to give an entirely different flavour. *Peppers* and *fennel*, of course, are also used in summer salads. For winter dishes, there are *winter radish*, giant forms with black skin, the form of chicory called *radicchio*, with leaves tinted, veined or speckled red, which will take the place of lettuce and can be grown outdoors or in a greenhouse, winter celery, and blanched chicons, as well as carrots and celeriac, grated or sliced, florets of cauliflower and red, white or green cabbage.

To fill that 'hungry gap' between early spring and early summer, sprouting broccoli, kale, leeks and spinach have already been mentioned; to them can be added radish, spring cabbage sown the summer before, a variety of lettuce which will stand the winter, sown in early autumn and cloched, carrots sown early in spring and covered, and in mild gardens, broad beans sown in late autumn also protected.

Unusual garden vegetables

The majority of the foregoing vegetables are common and familiar to most gardeners and gardens, though perhaps aubergines, courgettes, calabrese and sweet peppers are less so. Nevertheless, they are easily grown and are well worth trying, since they can be frozen in the form of ratatouille if there is a surplus. Fennel is another new flavour for late summer and autumn; so is *sweetcorn*, the latest varieties of which will produce cobs as early as the end of midsummer in cool temperate climates. *Jerusalem artichokes* are seldom seen, in spite of their ease of cultivation and unusually delicate flavour, perhaps because they can be a nuisance if not thoroughly dug out in winter. But they can serve a double duty as a windbreak as they grow to about 2½ metres (8ft) tall.

Celeriac and *kohlrabi* are two more good root crops seldom seen in shops and even less so in gardens; neither is

difficult to grow, both will provide roots in autumn and through the winter, so pay well for their place in a rotation. Swedes, for some reason, are rarely grown; their flavour can be more interesting than that of turnips and the weight of crop produced for a given area is greater. *Salsify* and *scorzonera* were once popular root vegetables whose tops can also be eaten in spring, so they provide two for the price (or space) of one. They take up little space and do not ask much in the way of soil or care.

Whether *pumpkins* and *squashes* are vegetables or fruit depends on the way in which they are served and the length of time for which they are stored, as the flavour gradually becomes sweeter with age. But the amount of crop they produce is huge compared to the space and time allocated to them, and their dual purpose makes them even more attractive. Some of the squashes will be ready for eating in summer and early autumn, others can be stored from mid autumn until mid-winter. Pumpkins can be extremely prolific and are said to have their cropping potential increased if planted amongst marrows (also called summer squashes or marrow squashes).

Colourful and dwarf varieities

As well as growing the straightforward varieties of each vegetable, that is, those which crop at the conventional and natural season and are the normally accepted colour and size, experimenting with new or unusual varieties can result in some unexpectedly pleasant surprises. The season of a vegetable may be prolonged; the new variety may suit your garden conditions better than the old one; the flavour may be different, even better; it may be much hardier or more prolific; it could be a different colour, so giving meals a more appetising appearance, and it could well be resistant to hitherto intractable diseases, and unacceptable to pests.

Golden-yellow forms of vegetables grow well and taste delicious; look for them amongst courgettes, sweet peppers and beetroot. Golden beetroot's advantage is that it does not discolour other food. There are yellow varieties of tomatoes, and one with red stripes on a yellow background; some are

Golden yellow courgettes (**left**) and white aubergines (**right**) are unusual varieties of these popular vegetables. Kohlrabi (**above**), an unusual root crop well worth growing, is available in two varieties, with either green or purple skins.

orange, and one or two are pink or white. Aubergines have many varieties which produce white egg-shaped fruit, much more widely grown in the tropics, but with the same flavour and no more difficult to grow in cooler climates. Celery comes in pink- and green-stalked forms, and there are lettuce tinted with red and Brussels sprouts flushed red, with red veining to the leaves and reddish stems. French beans can be purple-podded – they cook green – and climb (so getting them out of the way of slugs and snails) and so can peas. Onions can be red-skinned, cabbage can be white, cream or red.

There are small varieties of broad beans, runner beans, lettuce, tomatoes, cucumbers and Brussels sprouts. There are winter-hardy lettuce shaped like a cross between cos and cabbage with a specially nutty flavour, and lettuce which provide leaves continuously all summer. There are Japanese onions which stand through the winter and crop in early-late summer, and sweetcorn which matures much earlier than normal varieties. Triple-podded varieties of peas crop more heavily, and some potato varieties are particularly good for salads because of their flavour.

Look for parsnip which is resistant to canker, tomatoes resistant to leafmould, verticillium wilt and fusarium, and lettuce resistant to grey mould. Grow non-bolting varieties of beetroot, spinach and lettuce and use sets rather than seed of onions for the same reason. Much has been done by the plant-breeders to help both private and commercial gardeners and to ensure success in their vegetable cultivation.

Exotic vegetables

There is also a whole range of exotic vegetables appearing on meal-tables as a result of travel and the influence of ethnic cuisines. They are available, imported, in supermarkets and growing them in a cool termperate climate – some of them with the protection of cloches, tunnels or greenhouses – is possible.

Amongst these new exotic vegetables are okra (ladies' fingers or gumbo) chilli peppers, Hamburg parsley, Chinese artichoke, sweet potatoes, avocado, colokasi (taro) and yams.

The last three must have tropical climates before they will mature; the avocado is a fruit, only borne on a mature tree about 6 metres (20ft) tall needing at least sub-tropical climates to ensure continued growth. But the remainder will grow outdoors in warm termperate climates or cool ones with such protection as a greenhouse, frame or cloche, depending on their height.

Okra is grown for its seed-pods which follow a large and pretty flower of the mallow family; the pods are much used in Indian or Creole cookery. *Chillies* are easily grown in the same kind of conditions as sweet peppers; they are small and pointed with an exceedingly hot taste, not to be eaten or even sampled raw, but delicious in casseroles and curries. If they are dried and ground up, the resulting powder is cayenne pepper and chilli powder.

Hamburg parsley is totally hardy, and supplies a small but tasty root in winter, a cross between celery and parsnip in flavour, as well as parsley-flavoured leaves all through winter, useful when ordinary parsley succumbs to cold. *Chinese artichokes* are no more trouble to grow than their bigger relatives, but take up far less ground; *sweet potatoes* are another tuber which was commonly grown outdoors in the south of France at the end of the last century, and can be grown under cover in cooler areas, when they will store in winter in dry conditions above 4°C (40°F).

Some old-fashioned vegetables to try, to add still further to the range, could be *alexanders (Smyrnium olusatrum)* grown for its stalks and buds; *orach*, a leafy plant related to dock; *samphire*, with fleshy stems and leaves growing near to coasts; *Good King Henry*, another leafy plant; *corn salad* (lamb's lettuce) used in salads; *rampion* for its tuberous roots; and *skirret* which has a sweet taste.

It is sad that so many of these vegetables are not regularly grown and, even if space is lacking, the changes can be rung from year to year. In particular, it is worth searching out and reviving old varieties. Any garden will fit in a wider variety, with a bit of replanning, juggling and closer planting, and in the next chapter ways of doing this are suggested for different sites.

Vegetables in the garden

Suitable vegetables for the garden depend on a number of factors, one of which is personal or family preference. Peas, French beans, leeks and the uncommon vegetable, asparagus lettuce, are seen in this garden.

However small a garden, or however crowded with plants, room can always be made for vegetables in it somewhere.

Where no garden exists at all, any small patches of ground will be useful, but for many would-be vegetable gardeners, container-growing can be a satisfactory answer if they are city-dwellers with basement areas, balconies or roof-gardens. For those who only have window-ledges, there are still a few which can be grown, since there are vegetables specially bred for such conditions, or which are naturally small plants. For details of container-growing, see the following chapter; here we are concerned with open-ground cultivation.

Whatever area you have in which to grow vegetables, whether it is a large or small space set aside within the main garden, a patch of ground, an ornamental border where the vegetables have to be mixed in, or a greenhouse, there are some general planning considerations applicable to them all. It will pay to think about these in advance, in order to get the most from the space and soil available, and to get what you and the family actually want. In fact, family likes and dislikes must come first; it is no good growing a vegetable which is universally disliked.

Vegetables to suit soil and climatic conditions
It is also important to choose varieties which can be grown in the soil and the climatic conditions prevailing in the garden. Many soils will grow most vegetables, but there are some which are markedly heavy, with a tendency to be moist even in hot sunny weather and waterlogged in winter, and others which contain a good deal of sand, stones or gravel; these are 'hot', need heavy watering in summer and a good deal of feeding. Climate will influence the choice to a considerable degree; gardens where frost is likely will not allow the outdoor growing of tomatoes, aubergines, sweet peppers, marrows and so on in spring; lettuce, summer radish and broad beans would need winter protection, and so would carrots, beetroot and peas wanted in early spring. Even wind

can make a difference to successful growing, as insects are less likely to pollinate plants in windy gardens; and heavy rainfall with the dull light that necessarily goes with it are conditions best suited to the leafy crops.

What vegetables to grow?
What sort of vegetables are required? It may be a regular supply throughout the year, or it may be to supplement the choice of those which are cheap to buy in summer. Are flavour and nutrition the overriding factors in the selection? Should luxury crops, such as globe artichokes, asparagus and aubergines, take precedence or the unusual ones which are seldom grown and rarely available in shops – salsify, cardoon, Hamburg parsley or Florence fennel? What about vegetables to carry one through the time between early spring and early summer? If the space is short, choose varieties which are good value for the space they take up, and look for those which are dwarf forms. Consider your available storage space: garden shed, freezer, or spare ground for clamping. Finally, take your own available time into account; when plants need work of some kind done on them, it must be done at a particular stage in their growth; if the job is missed, eventually a whole crop may be lost. Vegetable growing is time-consuming, and requires time spent on it almost every day of every week, for most of the year.

Separate vegetable garden
The most obvious, easiest and most accessible way to grow vegetables is in a separate area within the main garden. Unfortunately for the rest of the garden plants, it should be in the most sheltered and warm part, with the best soil, and plenty of sun and no or little frost.

With an area allocated solely to vegetables, it is then easy to carry out a rotation to ensure that pests and diseases do not accumulate in the soil, and that mineral deficiencies do not occur because the same vegetables are grown in the same

section every year. It will also be easier to protect from birds and small mammals, and to have access for such chores as manuring, thinning, protecting from cold and staking.

Vegetables in small spaces
If the space set aside purely for vegetables is only a small one, adjustments to the techniques of growing have to be made accordingly. If a good variety of vegetables is wanted, there will be no opportunity for successive sowings of one vegetable such as lettuce. Spacing within rows and between them will have to be closer. Alternatively, patches rather than rows can be sown. There is no room for paths, although stepping-stones strategically placed, are advisable. Carefully choose your favourite vegetables, giving the best value for space – the soil will have to be particularly well fed and watered for intensive cropping of this kind. Cloches should be used as much as possible to get the crops in and out of the ground doubly quickly, and particularly careful thought should be given to the juxtaposition of plants, e.g. parsnips can entirely overshadow carrots, and Swiss chard will sprawl all over young endive when planted side by side. The practice of catch cropping is particularly important, when quickly maturing vegetables are grown in soil later covered by slower-growing plants, e.g. lettuce, radish or young carrots next to bush marrows, or ridge cucumbers.

Vegetables in greenhouses
Growing vegetables in greenhouses is largely applicable only to cool-temperate climates which require extra warmth and protection from weather in summer and winter. With a greenhouse, the vegetable gardener is able to grow satisfactory crops of tomatoes, sweet peppers, aubergines, indoor cucumbers, and okra; to start off vegetables not quite hardy in spring, such as sweetcorn, summer cabbage and summer cauliflower, potatoes, celery and celeriac, French and runner beans, cucurbits like marrow, outdoor cucumber, squashes and pumpkins; to force carrots, beetroot and radish; to provide protection for lettuce maturing in spring, and to do the same for red chicory to be harvested in winter, and to provide an area for blanching chicory, forcing rhubarb, and bringing on mushrooms.

In fact, the greenhouse should be in use for vegetable cropping all year round and is a tremendous asset for propagating as well. There is an option for the growing medium of the plants, because they can be planted directly into the soil of the greenhouse, or grown in containers. Direct planting will necessitate keeping the soil in excellent condition, with double digging, regular addition of well rotted organic matter, heavy watering in summer and, in most cases, liquid feeding as well. It is doubly important to rotate the crops in such a confined growing area, and to keep up the supply of nutrients. Ventilation is also vital to ensure a good circulation of air – the stagnant conditions that can prevail are an ideal breeding ground for fungus diseases.

Vegetable plots and allotments
Gardens which are too small to provide sufficient vegetables for the family needs, or which are too small to permit vegetable inclusion, can often be supplemented by acquiring a separate vegetable plot or, in the UK, an allotment.

For convenience's sake, it is best to obtain one as close to home as possible; some sort of shelter in which tools and machinery which can be locked will save a lot of fetching and carrying, and so will one which is close to a source of water. If this is not possible, a tank would be advisable. Choice of

Vegetable plots such as this one (**above**) in the French countryside can supply a family's needs all year round. Other vegetables like fennel (**left**) and certain varieties of cabbage (**above right**) can be grown for their decorative qualities, and where space is limited, vegetables can be fitted into flowering beds and borders (**below**).

plot or allotment should take into account the prevailing wind, as a screen may have to provided. If the land has been disused for some years, the weed problem will be considerable and can mean that a growing season is lost before it is fit for planting. New garden plots or allotments may consist of unsuitable soil. However, a lot can be done with regular feeding, working and manuring.

An area 10 × 3 metres (30 × 10ft) is ample to grow the family's vegetables, though it will not permit a full crop of potatoes sufficient for the whole winter unless other vegetables are sacrificed. It will certainly allow successions of varieties, planting in rows, paths, and room for a compost heap, also cultivation of the larger vegetables such as globe artichoke, or permanent crops like asparagus and rhubarb.

Vegetables in flowering beds and borders

Where the garden is so small that a separate vegetable bed is not possible, the vegetables will have to be fitted into the flowering beds and borders.

With flower borders, the decorative potential of vegetables can be exploited to the full. Small vegetables can grow at the front of the bed; cabbage lettuce make neat bright green plants and there is a variety with red tinting. 'Salad Bowl' lettuce have attractively frilled leaves, and so does the curly-leaved form of endive. The winter chicories are gaily and prominently marked with veins, variegations or flushes of red, and the fern-like foliage of carrots is graceful and airy. Beetroot leaves are a dark reddish green and French beans have red, white or pink flowers and yellow or purple pods as well as green ones. Dwarf broad beans have white flowers which are deliciously fragrant.

Decorative vegetables

Taller decorative plants include Florence fennel, whose filmy needle-like foliage clothes stems up to 90cm (3ft) tall. Asparagus is another airy plant. Ruby chard has bright deep red stems and leaf-stalks, red leaf veins and red tinting to the leaves; if it bolts, the handsome red flowering stems can be 75cm (2½ft) tall, but even low-growing it is a pretty plant. Globe artichokes are highly decorative, with their large, deeply cut, silvery-grey leaves and chunky flowerheads; one or two allowed to flower will be purple. The yellow variety of bush courgettes has large yellow flowers and not only golden fruits, but also leaves with patches of yellow; all marrows will have similar flowers. If aubergines can be grown outdoors, they are distinctly decorative with, in the purple-fruited varieties, large light purple flowers on deep purple stems, with leaves tinged purple. Sweet and hot peppers are white-flowered with green, red, golden and yellow fruits; tomatoes have clusters of yellow flowers and similarly coloured fruit.

Ornamental cabbages and kales are just that – extremely decorative and often used in winter flower arrangements, but just as edible as ordinary varieties. Colours can be pink, cream, purple, green, grey and white, with deeply frilled or serrated leaves, and in fact they look just like large open flowers lying on the ground.

Much taller plants for the back of the border or as a focal point in the centre could be runner beans whose red, white or salmon-coloured flowers were originally grown purely for ornament, or the purple variety of climbing French beans, with purple pods, purple stems, and deep purple flowers. Sweetcorn is a handsome foliage plant several feet tall, and for warm climates, okra has large pale yellow, red-centred flowers all the way up its 2-metre (6ft) stems. The leaf vegetables can provide cool greenery where the hotter flower colours of the ornamental plants are massed.

When mixing vegetables into the flower bed, care should be taken to see that they are not overshadowed by larger plants; it pays to plant in patches both for the appearance and for pest and disease control, particularly slugs. Only along the front edge can they be planted in a line. Extra care is needed when the plants are seedlings or small, and starting in soil blocks or 5cm (2in) pots is often advisable.

Vegetables in containers

Sweet peppers are ideal container vegetables and can add colour and ornamental value to patios, balconies and roof gardens.

Vegetables can be grown satisfactorily in containers, like any other plant. If there is no garden or plot to hand, or even a small section of soil within a paved area, the only practicable answer is to put the plants into their own individual gardens, i.e. a container with a suitable growing-medium. Basement areas, roof-gardens and paved patios are the largest suitable sites, limited only by cost of the containers and growing-medium. Most containers are long-lasting, so will be a capital cost, but growing-media need changing or refurbishing and will be a running cost.

Basement areas
Basement areas usually have some kind of hard surface such as paving, tiles, bricks or concrete. If it is soil, it is normally so poor and sour that it is not worth attempting vegetables, which need better soil than other garden plants. Other problems can include cold, because it is low down and cold air sinks to the lowest level; shade, unless it faces south when it may be almost too bright and hot; invasion by the local colony of cats, and a tendency for slugs, snails and woodlice to make it their base for a population explosion. Heavy rainfall can also lead to flooding. On the other hand, a below-ground area can be sheltered and warm; a well-lit one facing south-east, south or south-west could ensure the growing of tender vegetables, and any paved area is easy to keep clean and tidy, and therefore free from pest and disease.

Roof-gardens
Roof-gardens have considerable potential. The light is usually excellent unless there is a taller building nearby, and there is often a surprising amount of space. But winds can be fierce and damaging, even on days which are calm at street level, and screens are essential. Reinforced glass or synthetic glazing will cut down the wind, but not the light, without being too heavy, though sometimes there are already low balustrades in place whose height can be increased with securely fastened wooden trellis.

Weight is an important factor and can be minimised by using plastic containers and soilless composts. Even so, the plants need watering, and this adds considerably to the load-bearing demands made on the roof. Birds, especially pigeons and starlings, can be a nuisance in city roof-gardens, but other wildlife can include bees, butterflies and all sorts of flying insects. Pollination is no more problem than it is at ground level.

Balconies and window-ledges
Balconies and window-ledges outside apartments can be pressed into service, and large balconies can make a considerable contribution to a regular supply of fresh vegetables. Their problems are the same as basement areas, except that cats, flooding, and woodlice are less likely; birds and the weight-factor, as with roof gardens, are two drawbacks, and the lower occupants of the building, as well as passers-by, need to be remembered when watering or tidying.

Window-ledges or indoor window-sills can provide sites for troughs, pots and window-boxes containing low-growing or dwarf versions of vegetables. Ideally, outdoor containers in these situations should be fixed in some way with drainage trays beneath them; indoor ones should be in the lightest possible places with provision for some shade from the midday sun, also with drip trays.

Patios
Some small backyard gardens are completely paved. Container vegetable growing in these situations has no special problems and is limited by the size of the area, one's time and the money supply, all factors which have to be taken into account with traditional vegetable gardening. Sometimes it is necessary to grow crops in containers, even when there is a garden available, because the vegetable area is not large enough, and there are paved sites such as patios which will

take containers. If a greenhouse is too small for the variety of tender crops desired, pots can be put on its path at night, to be taken out into a sheltered, sunny place during the day.

Suitable vegetables for containers

The selection of vegetables to grow in containers will obviously not be as large as in the open garden, but with the right compost and sufficiently large or deep containers, a great deal can be done. Obvious kinds include aubergine, French beans, endive, leeks, lettuce, mushroom, onion and shallots, potatoes, radish, sweet and chilli peppers, Swiss chard, spinach, and tomato. There are also the dwarf varieties of broad beans and runner beans, a small variety of cucumber bred specially for containers, dwarf kale, Chinese artichokes, carrots, globe beetroot, kohlrabi, and turnips.

Greater depth of the order of at least 25cm (10in), and preferably more, is required for celeriac, the shortest versions of parsnips, winter radish, swede, and long or cylindrical carrots. Runner beans, climbing French beans and the tall varieties of peas need to be grown in tubs with a minimum diameter of 30cm (12in), and not more than two plants to a tub of that size. Cucumbers, marrows, courgette and squashes can be grown in grow-bags, but care is needed with the watering, as they have a tendency to wither-tip in these conditions, and a single container such as a tub for each plant will give better results.

Vegetables in the cabbage family such as sprouting broccoli, Brussels sprouts, cabbages and cauliflower, mostly grow rather large, and take space for a long time before they can be used, but there are dwarf versions of sprouts and cabbage if winter vegetables are a priority, and calabrese only grows about 60cm (24in) tall, though it crops in summer and autumn. Other vegetables which are really too large for container-growing are Jerusalem artichokes, pumpkins, sweetcorn, and three crops which have the added disadvantage of being permanent are asparagus, globe artichokes and rhubarb.

Chicory could be tried but the green forms develop quite a long tap-root, so need plenty of depth like the majority of the root-crops, and have to be blanched indoors in any case. Celery needs a lot of water and a very rich soil; the summer version crops relatively quickly, the winter one takes up room for six months and more, but either would be a bit of a gamble. Florence fennel is not an easy crop to grow in cool temperate climates in open ground; in containers it also needs a good depth, and plenty of water and warmth, but its size is suitable as it grows to about 45–60cm (18–24in) tall.

Types of containers

The variation in containers is considerable, both in size and shape, and it is a question of choosing what is suitable to the vegetables to be grown and the area available for housing the containers. The types of pots available are either clay or plastic. Clay are much heavier and breakable, but remain cooler and the walls of clay pots contain air and moisture. New clay pots need a good soaking for 24 hours before use otherwise moisture is sucked into the clay itself without benefitting the plant, no matter how much it is watered. Drainage material is required in the base. Plastic pots are easy to handle and do not break, although they can split with age. Soil-containing compost has a tendency to become hard and caked in them. They have more drainage holes and do not need drainage material.

Troughs and window-boxes are good homes for the smaller vegetables; they are seldom more than 15 or 17cm (6

or 7in) deep but a kind of row of vegetables can be grown in them and they are much better value for space than pots. Drums and plastic or wooden tubs are particularly good for climbing or larger vegetables, root crops or vegetables for exhibition. The extra depth would be rather a waste for small vegetables.

The proprietary grow-bag or plastic sack which already contains compost has proved its worth for tomatoes, sweet peppers, aubergines, lettuce, cucumbers, onions, cabbages, celery, dwarf French beans and small early beetroot, even sweet-corn, though this is a space-expensive crop. The bag is quite costly initially but it does include the compost, which can be used again the following season with a suitable compound fertiliser.

If the appearance of the vegetable is important, they can be contained in sculptured urns and jars of terra cotta which mostly provide greater depth than standard pots and tubs; strawberry barrels could be used, and plastic tower-pots are suitable for lettuce. The synthetic fibreglass troughs moulded and coloured to look like old lead troughs, with designs in relief on the outside, are another possibility, and are light in weight; the real thing would be just as acceptable to the plants. One ingenious arrangement of containers consists of five troughs attached one above the other to an A-frame shape, with the fifth trough on top of the A; it is economical of space and attractive. For vertical vegetables, there are now containers which can be stacked or hung against walls, containing separate pockets for vegetables such as lettuce, carrots, radish and other small root crops.

Tips for container growing

Whatever container is used, it should be at least 15cm (6in) deep and more if possible, except for grow-bags. Wooden containers must be treated with a plant-safe wood preservative and all should be completely clean before use. Provision for drainage, by means of drainage holes in the base, is essential, and all containers other than pots should be raised slightly off the ground to allow surplus water to drain away freely. A point to note here is that snails and slugs will use the under-surface as a hiding-place in daytime and for hibernating during winter. Containers glazed inside or outside are not suitable for plants.

Choosing a compost

Once the container has been chosen, the growing medium is most important and the proprietary mixtures and recipes are hard to beat. Satisfactory composts can be made at home but usually only after considerable experience. In the meanwhile, the soilless or peat-based kind containing peat and sand are excellent, provided regular liquid or foliar feeding is carried out according to the manufacturer's instructions. The soil-containing types, the John Innes composts in the UK, are also excellent as they are formulated for a great variety of plants and can be obtained as J.I. potting No 1, 2 and 3. The first is for plants of a size suitable for pots of up to 10cm (4in) diameter, the second for those up to about 17cm (7in), and the last for 20cm (8in) and over. In many cases, plants of the last-named size will not need extra feeding; it is the fruiting vegetables which often need a boost.

Although the container-gardener mostly has little space, if you want to make up the J.I. composts, the following is the formula for No 1 potting: 7 parts loam, 3 parts peat and 2 parts coarse sand, all parts by volume. To 36 litres (1 bushel) of this add 112g (4oz) of a mixture of 2 parts superphosphate, 2 parts hoof and horn and 1 part sulphate of potash, all parts by

Types of container Drums are suitable for root crops such as potatoes (**above**) which need some depth. Clay containers are commonly used for vegetables such as peppers (**left**) which do not need great height or depth, while compost-filled grow-bags (**right**) are worthwhile for a wide range of vegetables, including peas.

weight, together with 21g (¾oz) chalk. For No 2, double the fertilizers and chalk, and for No 3 triple them. A peat-based mixture is usually 75:25 granulated peat and silver sand respectively, to which a compound fertilizer can be added in quantities depending on the light of experience.

When choosing a compost, take into account weight considerations both for lifting and for load-bearing of balconies, window-ledges and roofs. Soil-based composts become short of organic matter as well as plant food, but if re-cycled can have soilless compost mixed with them, or rotted garden compost and a compound fertilizer. Access for transporting compost, perhaps in bulk, should be available – some roof-gardens can only be reached from a ladder through a trapdoor, or backyards only through the house.

Care of container vegetables

Vegetables in containers need a good deal of steady, careful watering from spring to autumn in most seasons, so watering-cans, hoses and a tap or tank are essential. In summer a container may need watering twice a day and provision should be made for watering while you are away on holiday, business and so on. Water, preferably soft, at atmospheric temperature is best and plants should be throughly watered and then left until the surface of the compost becomes dry again. Peat-based compost has a tendency to dry out suddenly and completely in a few hours; it can also become dry in the centre, even if the sides are kept moist by regular watering. Its weight is a good indication of the moisture

content and moisture meters are a great help.

Extra feeding is necessary for peat-based composts, using a proprietary liquid concentrate diluted as instructed by the makers and applied at the rates and intervals recommended by them. Soil-containing composts are less likely to run short but even so, it pays to feed tomatoes, aubergines, sweet peppers, courgettes and cucumbers. Liquid fertilizer can be applied to the roots or the leaves as a foliar feed but on the whole root-feeding is the commonest method.

Some plants need to be supported, and canes provide one of the most convenient ways of doing this. Rigid panels of plastic-covered wire trellis can be used, as can galvanised wire-netting and for grow-bags, special supports which circle the bag and provide a rigid vertical framework kept in place by the weight of the bag resting on its base. Where containers can be backed by a wall or fence, wires attached to these will enable plants to climb or twine up them, a good way of growing ridge cucumbers.

Watch should be kept on vegetables in containers particularly for: slugs, snails, greenfly, leaf suckers, whitefly and woodlice, amongst the insect pests: root-rotting fungus diseases brought on by waterlogging combined with pest injury to the roots; birds and cats; rapid drying out in excessive heat, or frost damage – wrap containers in lagging material in winter to protect the roots. Remove fallen and decaying leaves, flowers and fruits and dispose of the remains of the plants at the end of the season on a compost heap, built in a plastic dustbin.

Treatment of the soil

Double digging is frequently carried out to keep the soil in good condition. It is advisable between autumn and spring for a new vegetable bed or deeply rooting vegetables.

Vegetables are demanding plants because, although ornamental plants will always make some sort of display unless really badly neglected, vegetables will not crop satisfactorily unless the soil is in good order, or has regular and frequent attention. The soil in particular is important. If you do not start with a good, well-structured soil, or are not prepared to do the necessary work on a bad one, the vegetables grown in it will be small, tough, bitter and sparse.

Importance of fertile soil
A fertile soil, the kind which is known as a friable loam, crumbles easily in the hand, contains few stones whether small or large, is usually dark-coloured and has organic matter in it constantly rotting to produce humus. It also contains ample amounts of the minerals which constitute much of a plant's diet in a form which they can easily absorb. It will be slightly acid to slightly alkaline – extremes of either state lead to unavailability of the minerals, sour soil and much soil-borne fungal disease. A fertile soil will also be capable of releasing its water at a steady rate, so that sufficient is present for the plants' requirements and some is in reserve, without waterlogging or becoming parched. Finally, because of these good conditions, it will be a 'living' soil, containing worms, bacteria, and a variety of other fauna and flora.

Types of soil – clay-based and sandy
To produce such a soil would appear to be impossible, and certainly perfection can never be achieved but amazingly good results can be obtained without too much backache. A good soil balances on the line between too heavy and too light but for most soils the weight falls on one side or the other. Soil derived from clay is too heavy, that is, it retains a great deal of water which literally makes it weigh more, leading to lack of aeration and therefore poor root functioning. It discourages the presence of soil-living organisms, and prevents the proper decay of organic matter so that the soil moisture gradually accumulates toxic quantities of salts.

Soil containing a good deal of sand is well-aerated but loses its water rapidly, and with it goes the nutrient content. At the same time, because such a soil warms up quickly, any organic matter present rapidly burns away so that it needs to be repeatedly and expensively fed with such material. Such a soil, like the clay-based kinds, will also be deficient in soil organisms, though it does at least have the merit of getting warm quickly in spring, thus making it easier to produce early vegetables, and it is also excellent for germinating seeds. But a sandy soil is not called a hungry soil for nothing, and again, its opposite does have another merit: a heavy soil can be extremely fertile because of the nature of the soil particles, and when well-worked – dug and manured, mulched and cropped for some years – its mineral content becomes available to the plants; the particles of a sandy soil do not inherently have this feature.

The difference between the two types of soil is easily distinguished; moist clay soil will feel sticky in the hand, will crack badly in dry weather, and any holes dug in it will remain full of water after heavy rain for hours or even days. Sandy soil has a distinctly gritty feel when moist, and will always be easy to dig even after heavy rain; it does not crack but simply becomes hard and smooth on the surface.

Adding organic matter
The best method of restoring or improving soil structure is by adding rotted organic matter, of which there are many kinds. As a buffer between extremes of water content and pH values, it works very well whether the soil is fundamentally clay- or sand-derived and, as it rots still further to complete the process of decay, it will supply humus, a finely-divided black substance. This will act rather like a sponge absorbing a great deal of water and yet still containing air.

Organic matter consists of such materials as farm dung, whether from pigsty, stable or cowshed, home-made garden compost, leafmould, seaweed, peat, spent mushroom compost and deep litter from intensively-kept hens. The farm

manure should never be used fresh – it should always be well on in the rotting process; seaweed can be used fresh, though it is better to compost it, either alone or in the general compost heap. Peat contains virtually no nutrients and plant food must be added if it is used on sandy soils. Spent mushroom compost can be used at once; if it has an alkaline reaction, care is needed for repeated use. Straw-based deep-litter can be dug in at once, but shavings-based litter needs to be rotted for up to two years.

Quantities to apply are about 2.5–5kg per sq metre (5–10lb per sq yd) but it depends a great deal on the state of the soil and how bad its condition is, and to some extent on the type of organic matter. The lower rate is better for seaweed, the higher for peat. For really sandy soils, a high rate like 8.5kg (18lb) would be preferable, but where a soil is in good condition, the low rate of 2.5kg (5lb) will be suitable.

Any of these should be evenly spread over the soil surface and mixed in while digging in autumn or winter; they can also be applied as surface dressings or mulches, along the rows and round individual plants, remembering that roots spread as far underground as does the top growth above it.

Soil cultivation
The soil can also be kept in good condition by cultivations, of which the chief is winter digging. Single digging, with a spade, is digging to the depth of the length of the blade; double digging is two blades' depth with forking of the base of the hole or trench so produced. Both are called single and double spit digging. By loosening the soil, aeration and drainage are improved, humus can be added and the remains of crops and weeds removed. For a new vegetable bed, for deeply rooting vegetables and show vegetables, and for heavy soil, double-digging is advisable between autumn and early spring; in summer and for other crops, single digging or forking will generally be sufficient.

Forking is mostly done where the soil is already reasonably loose or no great depth is required, and in summer where one crop is to follow another. Forks can also be used to prick the soil surface and help aeration in that way. Hoeing is mainly a method of removing weeds and creating a dust mulch to prevent moisture evaporation in hot weather. Raking is hardly a cultivation, more a way of creating a fine texture on the surface, as is required for seed-beds.

Fertilizers
While bulky organic matter is vital for good soil structure, initially it does not add much plant food, though as time goes on, this will increase in the soil with its gradual improvement. Extra plant food can be given as compound fertilizers, products which contain the three most important minerals: nitrogen, phosphorus and potassium, and sometimes also the trace elements, essential minerals required in minute amounts. Nitrogen helps with leafy crops, phosphorus with seedlings and maturity, and potassium with water content, cell metabolism and flower/fruit production. The trace elements have a variety of functions, and deficiencies of any of these will produce different symptoms in the plants, such as whiptail in cauliflowers due to lack of boron. In the average garden, however trace element deficiencies are rare, and the quantities required are generally present in organic matter in any case.

Plants absorb compound fertilizers in solution through their roots, so in dry weather they will be short of food as well as water; liquid feeding ensures that the minerals are absorbed quickly; liquid foliar food (leaf-feeding) is even more rapidly absorbed. Powder fertilizers sprinkled on to the soil need to be watered in, or better still rained in, but whatever the method, they should be applied to soil already moist.

Maintaining a balanced pH
Reference has been made to the pH of the soil. This is a measure of soil's acidity or alkalinity. It is obtained by mixing the soil with water, allowing the solution to clear and then testing this with a reactive chemical. The pH scale ranges from 1–14, with 7.0 as the neutral point; the lower figures are acid, the upper ones alkaline. In extreme acidity (4.0 or less) or extreme alkalinity (8.5 or more), the chemistry of the soil changes so that minerals become unavailable, or available only in toxic quantities. It is important, therefore, to maintain more moderate levels and regular applications of organic matter will help to do this; if alkalinity is required where the soil is naturally acid, lime can be used to alter it.

Chalk (calcium carbonate), ground limestone (limestone flour) – a similar material but slower-acting, and magnesium limestone can be used; rates vary according to the soil pH value, and soil-testing kits will advise about suitable quantities to use. Never lime indiscriminately; if an alkaline soil is made more alkaline, it will be very difficult to alter and can make conditions quite unacceptable for the plants. Test the soil first to see if it requires lime, and then add it during the winter, with a six-week interval between its application and that of organic matter; never add at the same time as fertilizers.

Organic gardening
While fertilizer and lime application are the standard methods of feeding the soil and helping to maintain a balanced pH, there is the organic school of thought, which maintains that plants will be stronger, healthier, and in the case of crops, better to eat, if grown without the use of 'chemicals'. This is a rather loose term used as an umbrella for concentrated manufactured fertilizers, insecticides, fungicides and herbicides; ground minerals such as rock phosphate and the limes mentioned are not manufactured in the same sense but are the raw material reduced to particles similar to those naturally occurring in the soil.

There are organic fertilizers, such as bonemeal, to supply phosphorus; hoof and horn meal or dried blood to supply nitrogen, and wood ashes to provide potassium. These mostly work more slowly but are longer-lasting than the inorganic kind; they are much less likely to be washed through the soil beyond the plants' reach by heavy rain, so are more economical in the long run. In vegetable gardens where the planting is intensive, small quantities of these may well be needed in addition to organic matter, especially on quickly-draining soil.

Organic gardeners maintain that crops grown with organic manures and fertilizers are healthier and taste much better and, that in the long run, it is cheaper to garden this way. Organic treatment of the soil means always adding organic matter every season to one part of the bed or another, and using the slow-acting fertilizers derived directly from animals, birds, or fish, or from seaweed and ground minerals.

One specialist organic school maintains that digging or any cultivation is harmful and unnecessary. Provided a good layer of rotted organic matter is spread completely over the soil and renewed as necessary – therefore forming a permanent mulch – it will be naturally absorbed into the soil by the action of weather and soil fauna, particularly worms. As a result the soil's structure will be maintained and improved to increasingly greater depths. Such a method also helps to prevent and suppress weeds, protects the soil from frost, and keeps it moist in summer. But whether you are organic or inorganic, mulches will always be important and it is always a problem to find enough material to treat the soil.

Availability of organic matter

Farm manure is rarely available nowadays; not everyone lives near a mushroom farm or the coast, and peat is expensive. Proprietary forms of battery-poultry and farm manure treated to produce a stable, clean, non-odorous product can be obtained but are also expensive; leafmould and garden compost can both be made at home and cost nothing except one's time, and not much of that.

Garden compost consists of a mixture of vegetation, rotting down in a heap until it turns into a dark brown-black, moist, crumbly material. It is made of waste vegetation, such as leaves, grass mowings, weeds, roots, soft stems, flowers and fruit. Hard material like woody stems, bark, leathery leaves, fruit stones and tough stems of the cabbage-stalk type are not used, as they take years rather than weeks to decay. Waste material from the kitchen can perfectly well be used as well, provided it is vegetative: potato peelings, apple parings, tea-leaves and coffee grounds, outside cabbage and lettuce leaves and so on. If a compost heap is a technique new to you, do not put in the roots or seeds of the more pernicious weeds, as the heap may not heat up sufficiently to kill and rot them, and try to avoid diseased material for the same reason. Even when expert at making garden compost, never add any of the cabbage family that has been infected with club-root disease – the disease will not be killed and it would be disastrous to spread it further over the garden.

Making your own compost heap

Whatever materials are included, there are a few simple principles to observe in building the heap. If it can be regarded as a kind of fire, the correct method of making it will be easy to follow, since though draught is essential to keep a fire alight, and similarly with a compost heap, air should be present, preferably entering from beneath it at the base and going up through it. If grass cuttings are put into the heap without being diluted, they will rot to a certain point and then turn into a slimy, unpleasant-smelling mat of half-decayed vegetation, due to lack of air and the retention of too great acidity. The base can consist of bricks or stones in lines with tough prunings, twigs and shoots laid across them, to keep the lowest layer just clear of the soil. Put several stakes upright within the area to be covered by the heap, and build it round them; when it is finished, remove them, thus ensuring air-shafts down to the base.

Besides the presence of air, there should be moisture, but not too much. A wet heap prevents warming up and therefore decay; dry heaps also cease to rot and provide homes instead for woodlice, mice and rats. The natural moisture present is sufficient with summer rains, but a cover is advisable for winter heaps and to keep out the heaviest downpours.

There should also be nitrogen present, provided by the green material, and the heap should not become markedly acid, so lime needs to be added. Warmth is also necessary, but this is suppplied as the fresh green vegetation rots and, provided it heats up to a high enough temperature, about 82°C (180°F), weed seeds will be killed. All these factors: air, moisture, nitrogen, heat and a moderate pH value provide the ideal environment for bacteria, fungi, worms and other living organisms to feed on the heap and convert it to the right state and substance.

Heaps should be about 120cm (4ft) high and 150cm (5ft) square for the best results, though smaller ones will provide good material from the centre. They are built up in layers, 20–23cm (8–9in) deep, of vegetable matter, with a thin layer of animal manure if possible, alternating with a sprinkled layer of lime. The base should consist of coarse material as already described. Containers can be made of wood or plastic, not wire-netting, as the outside becomes dry, and it is difficult to get at the material as it becomes inevitably embedded in the wire netting. For small quantities, use plastic dustbins, plastic sheet or plastic sacks, but make provision for aeration.

Making leafmould is simple; use any leaf except tough or leathery ones, do not include twigs and sticks. Pile the leaves in a heap as for compost, and leave to rot down for at least a year. Turn the sides to the middle after six months or so.

Soil testing kit used to measure pH levels in the soil.

Starting vegetables

Vegetables grown from tubers/bulbs
ARTICHOKE, JERUSALEM
ONION
POTATO

Vegetables grown from roots, suckers, young plants
ASPARAGUS
ARTICHOKE, GLOBE
RHUBARB
SALSIFY
SCORZONERA

Vegetables grown from transplants
BRUSSELS SPROUTS
CABBAGE
CALABRESE
CAULIFLOWER
LEEK
SPROUTING BROCCOLI

Using a trowel and garden line to plant out cos lettuce in early spring.

The care with which vegetables are looked after when their lives start determines their future health, strength and cropping potential. Even if weather conditions are savage later on, a plant with a strong constitution will survive and mature to some degree, whereas a weak one, perhaps checked as a seedling through lack of water, or planted with a kinked stem, will more than likely die. To ensure a strong seedling, preparation of the soil for seed must be thorough, and if an indoor start is necessary, a special compost for seeds should be used. Vegetables grown from plants do not make such demands on soil preparation but do require careful planting, whether a young plant, a transplant, a division or a tuber.

Growing from seed

The majority of vegetables are grown from seed and are sown outdoors in spring, even in cool temperate climates. For successful germination, a seed must have water, oxygen, a rise in temperature, and nutrient. Some seeds can germinate without water but will die if they cannot absorb some soon afterwards; some need considerable warmth to break their dormancy, others will sprout in quite cold temperatures. Most seeds contain some food for the sprouting rootlet and shoot, beans and peas in particular, but it is essential that phosphorus is available to the roots, since much of what the plant needs is acquired while it is a seedling.

Outdoor sowing

In order that a seed-bed can provide all these conditions, it will need forking over to break down large lumps of soil; at the same time all weeds and large stones should be thoroughly removed. A general compound fertilizer can be added now or about ten days before sowing, containing a higher proportion of phosphorus, or equal amounts of the three major elements, at the rates recommended by the manufacturers, or at half the rate, giving the remainder half way through the life of the crop. This can be done in advance of the final preparation which will be on or the day before sowing.

The time to sow should be chosen carefully when the seasonal temperature is obviously rising after winter, and when the soil is dry enough to rake without sticking to the teeth of the rake. Sowing in soil which is cold, or fairly wet, with the prospect of further cold or wet weather to come, will ensure that seeds die at once, or rot, if they do germinate. If the season is late, cloches or tunnels put over the forked soil will help it to dry and warm up.

A firm, level seed-bed is necessary to avoid waterlogging in hollows, and seeds being sown too deeply for germination. Treading it first will firm it adequately, then rake twice, crosswise the second time, to produce a crumb-like texture to a depth of about 15mm (½in), though this is not essential for large seeds like beans. The seed-bed will usually be where the vegetables are to be grown, but some are started in a nursery-bed and transplanted from it to the vegetable bed proper.

Methods of outdoor sowing

Outdoor sowing can take the form of one of three methods: in rows, in 'stations' and in patches. Rows are the accepted method and are known as drills; they are V-shaped and made with the corner of a hoe drawn along the soil, guided by a length of taut line attached at each end to stakes. Depth of the drill should be even, with no small stones or lumps of soil in it.

'Station' sowing consists of putting two or three seeds together at evenly spaced distances in the drill, rather than a complete line, and patch sowing means to cover small irregularly-shaped areas with seeds evenly spaced all over it; rather than remove soil, it is easier to put a suitably deep covering over the seeds after sowing.

Sometimes, in order to help accelerate germination, fluid sowing is carried out. The seeds are contained in a colourless jelly-like paste, which is then squeezed through a

Vegetables grown from pulse seeds
BEAN, broad/French/runner
PEA

Tender vegeables planted out
AUBERGINE
BEAN, French/runner
CELERIAC
CELERY
CUCUMBER
MARROW
PEPPER, SWEET
POTATO
PUMPKIN
SQUASH
SWEETCORN
TOMATO

Vegetables grown from seed and their viability

Name	Years
AUBERGINE	6
BEAN (all)	2
BEETROOT	4
BRUSSELS SPROUTS	4
CABBAGE (inc. Chinese)	4
CALABRESE	4
CARROT	4
CAULIFLOWER	4
CELERIAC	6
CELERY	6
CHICORY	6
COURGETTE see MARROW	
CUCUMBER	7
ENDIVE	5
FENNEL	3
KALE	4
KOHLRABI	4
LEEK	4
LETTUCE	3-4
MARROW	7
MUSHROOM (dry spawn)	6 months
OKRA	4
ONION	1-2
PARSNIP	1
PEA	2
PEPPER, SWEET	4
PUMPKIN	4
RADISH	4
SALSIFY	2
SCORZONERA	1
SPINACH	2
SPINACH PERPETUAL	4
SPROUTING BROCCOLI	4
SQUASH	7
SWEDE	2
SWEETCORN	2
SWISS CHARD	4
TOMATO	3
TURNIP	2
ZUCCHINI, see MARROW	

funnel along the line of the drill; the paste can be a cellulose-based wallpaper paste without a fungicide, mixed at half-strength so that the seeds do not sink. A plastic bag can be used as a funnel if a corner is cut off and the jelly sqeezed through it. Depth of sowing is about 6–15mm (¼–½in), depending on the size of seed, and is roughly twice that of the diameter of the seed. Water the base of the drill or patch before sowing and, if the soil structure is not very good, line the drill with granulated moist peat.

Care after outdoor sowing

Once the seeds are sown, cover with fine, crumbly soil and firm down lightly; protect from birds and cats. If no rain occurs, keep the surface soil moist with regular light watering. Remove any weed seedlings, and continue to do so after germination. In cold conditions, replace cloches until the weather improves, even after germination. If the seeds have been sown where they are to grow, most will need thinning, and this should be done when they are large enough to handle to leave about 25mm (1in) between the seedlings. A second thinning to the final spacing can be done when the leaves touch in the row. Thinning should be done as soon as it is necessary, otherwise seedlings become weedy, and the soil should be firmed back around those retained. In a nursery-bed, seedlings are usually thinned to about 10cm (4in) apart.

Indoor sowing

Vegetables which are vulnerable to frost or which need temperatures of 16°C (60°F) in which to germinate must be sown under cover in cool temperate climates. They are sown in containers of seed-compost, pricked out or potted on still under cover, and then planted in their permanent sites either in a greenhouse or outdoors in a warm, sheltered, sunny place.

Preparing the seed-tray

The standard method is to fill a seed-tray (flat) with compost and sprinkle seed evenly on the surface. Sowing a few seeds in individual containers and thinning to one seedling, as with station-sowing, allows more specialized attention, and avoids pricking-out and the root disturbance that goes with it. A seed-tray can be plastic or wooden; the latter will need a drainage layer. A proprietary peat-based compost or soil-containing seed compost should be used, and if the latter is to be made at home the formula for the British J.I. version is: 2 parts loam, 1 part granulated peat, 1 part coarse sand, all parts by volume, to 36 litres (1 bushel) of which mixture is added 42g (1½oz) superphosphate and 21g (¾oz) chalk or ground limestone, thoroughly mixed in. Peat-based potting composts are commonly used for seeds as well.

The tray should be filled to within 15mm (½in) of the rim, and well firmed down if it is J.I. compost, lightly firmed if peat-based. The firming is best done with the fingers, starting at the corners and sides, and finishing in the middle. The surface is levelled off and firmed with a board, and the tray placed in a shallow container, such as a domestic tray, in which there is water, and left there until the surface is dark and moist without the base being soggy; this only takes a few minutes.

When sowing the seed, try to space it evenly on the compost, then cover with moist compost, sieved if necessary through a fine mesh, firm down lightly and cover with black plastic sheet. Temperatures for germination should be about 16°–18°C (60–65°F), up to 21°C (70°F) for cucurbits and okra.

After germination, remove the plastic sheet, give the seedlings plenty of light without scorching, keep them at about the same temperature, slightly lower at night, and keep them moist by watering with a fine spray.

Pricking out

When the first true leaf appears after the initial seed-leaves, move them into another tray or into individual 5cm (2in) containers. This is called pricking out. It must be done with as little damage to the roots as possible, handling the seedlings by the stems, and lowering them into a hole which ensures that the roots are not doubled up. The stem of the seedling should be buried so that the seed-leaves are only just above the compost surface, then the compost firmed gently round the seedling, and the tray watered when planting is complete. Spacing is about 5cm (2in) each way, closer for small seedlings, and an average tray will take 25–30 of them. After pricking out, the seedlings should be put in the shade for a day or two. When they have grown so that the leaves of neighbouring plants are touching they are ready for planting out. However, they are hardened off first, that is, accustomed to outside temperatures gradually by putting them outdoors during the day to start with, and then in a frame with the top closed at night, until the top can be left slightly open.

Starting seeds in individual containers

Seeds sown in individual containers from the start are more trouble but produce better plants, and it is often the only way of side-stepping slug and snail trouble, as well as avoiding root disturbance during pricking out and planting. It is also easier where vegetables are to be grown permanently in containers. Individual starter containers can be 5cm (2in) plastic or clay pots, peat pots or pyramids, compressed dry discs of peat contained in a fine plastic mesh which swell to three times their size when wet; propapots, a series of plastic pots moulded all in one with a tray; and soil blocks made of compressed compost.

Potting on

Where vegetables have to be transferred from starter pot to growing pot, and then through larger pots into the final pot size, the stage for potting on in each case is when the roots fill the soil-ball and are just appearing over the outside of it. If they have reached the point of coiling round the base, potting is overdue.

Planting

The majority of vegetables are grown from seed, but there are some which are grown from plants, resulting from the method of increase by rooted suckers or divisions, and from tubers or bulbs. There are also the young plants transplanted from a seed-bed, or plants of tender vegetables which have had to be brought on under cover.

Preparing the soil for planting

The soil does not need such intensive preparation as it does for the production of a seed-bed; nevertheless for most of them it needs to be single-dug and for some, particularly the permanent crops, double-digging is preferable. But, provided the surface is broken up so that large clods are removed, together with large stones and weeds, thus making the surface reasonably friable, no more need be done except to give a dressing of a concentrated compound fertilizer, if the vegetable is one that will benefit or if the soil needs it. One of the general kind will be adequate but if you want to

Left Using a draw hoe to form drills suitable for sowing beans. **Right** Planting sprouted seed potatoes in a drill. Potatoes need to be positioned well apart since the top growth takes up a lot of space.

be more precise, one with a high nitrogen percentage for a leafy vegetable, or a high potassium or phosphorus content for a fruiting crop, can be applied. This should be evenly spread about 10 to 14 days in advance, to allow sufficient time for it to be absorbed by the soil before planting, and it should be forked into the top few cm (in) of soil.

When to plant

The time to plant can be between late winter and early autumn; Jerusalem artichokes and shallots go in during late winter, globe artichokes can be planted in mid spring or early autumn, Brussels sprouts and leeks transplanted in early summer. Whatever the time, plant when the soil is crumbly and easy to work, when there is a likelihood of rain to follow soon after. In the summer, plant in the evening and water in immediately afterwards.

Spring planting will get off to a quicker start if planted in rising temperatures and in soil that has been warmed in advance with cloches placed on it for a fortnight. In summer, the problem is more likely to be high temperatures, and watering is the key to a successful 'take', both at the time of planting and every day in the evening until rain occurs. The moist conditions in otherwise dry areas are likely to attract snails and slugs, and precautions against them will inevitably be necessary.

It goes without saying that for these vegetables, the planting hole should be wide enough and deep enough to allow the roots their natural spread; also ensure that the soil mark on the stem, or the crown itself, is on a level with the soil surface. Firm planting is important. Care is necessary with potatoes to avoid injury to the young shoots; onion sets are only partially buried, and the tubers and bulbs in general rarely need watering in.

Transplanting vegetables

All the cabbage family are transplanted, and once they have reached the stage of about four good leaves in the seed-bed, they are ready to be moved. Plants which have no growing tip or which have kinked stems are not worth transplanting. Both the plants and the permanent site should be well watered the night before; the plants are dug up, using a trowel, with as much soil as possible on the roots. The hole is made for them with a dibber, a short wooden pointed tool with a handle about 25mm (1in) in diameter. The hole should be filled with water and allowed to drain away, then the plant put in so that its lowest leaves are only just above the soil surface; it should then be firmed and watered in. Celery virtually needs to be flooded; it is a water-loving crop, which more than most should never run short, hence the need for plenty of rotted organic matter to provide a reserve.

Leeks are ready for transplanting when they are about 15–20cm (6–8in) tall. They are also planted in holes made with a dibber, about 10–15cm (4–6in) deep, and simply dropped into them so that they lean against one side. The roots can be trimmed back to a reasonable length, and the tips of the leaves cut off, mainly to prevent them trailing on the ground and being drawn into the soil by passing worms. Each hole is then filled with water and within a matter of days the plants will have taken root and started to stand upright; the hole will gradually fill with soil as time goes on.

Tender vegetables planted out

In cool temperate climates and even some warmer ones, there are some vegetables which have to be sown and grown under cover until it is warm enough to plant them outdoors or in a greenhouse. Such plants must be hardened off and, in particular, accustomed to lower night temperatures before planting. It is also important that they have not become starved in their containers. If the outdoor weather continues cold and unsuitable beyond the normal time, container-grown vegetables should be liquid-fed at regular intervals. It may even be necessary to give them larger, 10cm (4in) or more, containers to tide them over, especially if they are the vigorously growing marrow family. Tomatoes, aubergines and sweet peppers will be about 12–15cm (5–6in) tall, marrows and their relations will have three–four leaves when ready for planting.

When planting, water the plants well the evening before, and water the planting hole. Make sure it is large enough to take the soil-ball, and plant without disturbing the roots more than can be helped, unless the plant has become unavoidably pot-bound. If so, cut back the long roots, and gently ease the soil at the sides of the root-ball so that it is not a solid compacted mass into which air cannot penetrate, otherwise the plant will never grow new roots and water will not penetrate the root-ball.

Planting should be at the usual level, so that the stem soil mark is at the lip of the hole, followed by watering in. Some vegetables, such as French or runner beans and tomatoes, need staking, and any supports should be provided before the plant is put into the hole, to avoid disturbance and injury to the roots.

Care while growing

Once most vegetables are under way and have reached the small-plant stage (after thinning, transplanting or final potting), there is not a great deal of actual work to be done, although they should be looked at regularly and frequently. However, it is essential that what there is should be dealt with at the appropriate time, and not skimped or even missed altogether. Small, but regular attentions provide the good, heavy crops that are the hallmark of the professional.

Dealing with weeds

There is much talk about the complete removal of weeds at all times and the necessity for ensuring that the vegetable garden is permanently free of them. Weeds between the rows do make them look untidy, and from the time when vegetables are germinating to the young plant stage, weeds germinating with them should be removed as soon as possible. Weeds grow more vigorously than the crop plants and absorb much-needed water and food that were intended for them. They can thus weaken the vegetable plant, slow down growth and eventually produce a poor adult plant.

As the vegetables are germinating, hand-weeding between them may be the only possible method, especially for leeks and onions; as they grow, hoeing with an onion, Dutch or draw hoe will probably be possible, and later in the season, hoeing will also help provide a dust mulch to prevent moisture evaporating in hot weather. Alternatively, a chemical weed-killer can be used which will kill weeds beneath the soil surface as they germinate without harming the crop plants; one such is propachlor.

Hoeing weeds out while they are seedlings is always the best policy, but sometimes it is too wet to keep up a regular hoeing programme. If they grow larger in the later stages of the crop's life, it is not too detrimental, provided the tops are broken off before they can flower and set seed. In fact, some weed cover can be helpful in hot weather, as it actually helps to keep the soil moist, though a proper mulch would be preferable.

Perennial weeds such as bindweed, couch-grass, ground-elder and dock should not be allowed to establish, particularly amongst asparagus, as they are so difficult to eradicate. Glyphosate sprayed or painted carefully only on to the top growth of weeds is effective and avoids harming the vegetables if kept off their top growth; and for annual weeds like chickweed and bittercress, paraquat can be used, again keeping the spray off the vegetable top growth. Weeds will be at their worst in spring, but watch for their stealthy incursion in autumn, and clear the ground of them thoroughly before winter, otherwise what appeared to be a few tiny seedlings will be a mat of tough cover by spring.

Applying a mulch

While weeds can act as a kind of emergency mulch, the normal mulching materials, if nothing else, act as weed-suppressants. An organic mulch, such as rotting garden compost and similar substances, will help the soil to keep its moisture (especially helpful in dry or hot weather), will supply some plant foods slowly and steadily throughout the crop's life, and will maintain the soil structure. With so many advantages, it seems a short-sighted policy not to use a mulch of this kind.

Application should be when the plants have settled down after thinning or transplanting, and after any application of fertilizer when the soil is reasonably moist. It can either be spread all over the bed, or put on directly along each side of a row leaving any paths clear; a useful mulch for paths can be lawn mowings, applied as the grass is cut each week. It prevents the soil from compacting too much and will rot into it in time, or can be dug in at the end of the season, as any remains of the bulky mulch are. Thickness of the mulch can be 1.5–2.5cm (½–1in) or more, depending on the type of soil, although in practice spreading it to an even thickness is not easy, and it often finishes up in a rather lumpy state. Soils which are in good condition need only be mulched every year, unless a crop is a greedy feeder, such as beans.

Need for moisture

One of a mulch's most important functions is to conserve moisture, and summer droughts can cause a lot of problems. One of these is running up to flower and setting seed, known as bolting, and happens with biennial vegetables such as beetroot, or lettuce. In both cases the plants' strength goes into the flowering and it does not form a good root or a good leafy heart; onions and ruby chard are also offenders of this type. Dry weather in late spring, not necessarily hot, is another danger time. Hot weather in early autumn is not quite so crucial as many vegetables will have completed their growth and be maturing and ripening. Windy dry conditions are the worst of all, as wind can cause such rapid transpiration as to result in leaves browning and withering.

Artificial watering is almost always necessary to ensure decently-sized plants; shortage of moisture leads to cracked, small, tough root crops, bolting and bitter-tasting leafy vegetables, non-setting of beans and fruiting crops, small curds in cauliflower and calabrese, small bulbs in onions, shallots and fennel and probably bolting as well, and a rash of insect pests – greenfly, blackfly, whitefly, red spider mite and root aphis. Heavy rain or watering after drought results in splitting and cracking of many fruiting vegetables, potatoes and roots. Hence, watering should be started long before it is apparently necessary. Even if the soil is well saturated and a week of dry weather follows, watering should be started and maintained until rain occurs again.

Water requirements of different vegetables

Every crop has different water requirements, and the refinements of quantities and times of application are too great for the space available here for discussion. However, as regards the types of vegetables in general, the following recommendations can be made. Fruiting vegetables, i.e. tomatoes, aubergines, sweet and chilli peppers, marrows, cucumbers, squashes and pumpkins, peas and beans, and sweetcorn, should be watered, if drought conditions are occurring, once at flowering and once as fruit begins to swell, at about 9 litres (2 gal) per sq metre (sq yd). If there is water available and dry conditions continue, watering should be twice a week, and daily for tomatoes in greenhouses.

Root crops are less demanding and need only be watered fortnightly, but with at least 16 litres per sq metre (3 gal per sq yd), preferably more but with less while young, about 4.5 litres (1 gal). Leafy vegetables do best with about 16 litres per sq metre (3 gal per sq yd) every week. Onions rarely need watering except in extreme conditions of drought, and even so, not as they approach maturity. Early potatoes will give good yields with 16 litres (3 gal) applied fortnightly from mid-spring; maincrops can be watered heavily with benefit once flowering starts, at 22–27 litres (4–5 gal).

In every case, it is better to apply plenty of water occasionally rather than daily dribbles. Use a hose or watering-can without a spray close to the plants, but without such pressure as to wash the soil away or to force them out of the ground. One last point: it is worth remembering that vegetables grown slightly on the dry side will have more flavour – pumped up with too much water, they can end up large and tasteless.

Supporting plants

During the growing season, some vegetables will need to be supported. Obvious crops are such naturally climbing plants as peas, beans and cucumbers, and there are others with upright growth which needs supporting, although they do not

have particularly long shoots.

Runner beans, and climbing French beans need strong, tall stakes or poles, securely fastened at the ends of rows. They can be arranged as a double line of slanting poles sloping towards one another to form inverted Vs and crossing at the top with a horizontal pole lashed along the row at the junction of each pair of poles. Other supports can be a straight line of poles, less secure because it presents an unbroken, upright line of foliage; wigwams of poles in a circle with the tops meeting and tied together; and a central stake with arms radiating from it with strings hanging down to ground level, one for each plant.

Ridge cucumbers can be trained up wire-netting with a wide mesh, rather than trailing along the ground; trailing marrows grown for courgettes can be similarly treated. Indoor cucumbers will need strong supports such as wires attached to the inside of the greenhouse in a grid pattern. Container-grown plants can have a single stake each, but sideways growth has to be very restricted.

Cordon tomatoes can also be attached to single canes, or twisted round soft string attached at the base to a wire hook in the soil or to a horizontal wire, and to a hook or wire at the top. Brussels sprouts and sprouting broccoli often need a single supporting cane, and French beans need short brushwood to keep them from trailing on the soil. Dwarf peas also can be supported with brushwood or two rows of wire netting; broad beans simply need two lengths of wire, one on each side of the row near the top of the plants, attached to stout stakes at each end.

Protection against birds and small mammals

Protection will be required for many vegetables at some time during their life against birds, especially sparrows, pigeons and blackbirds, against small mammals – mice, voles, moles, rabbits – and against cold. Sparrows and other small birds scuffle about on seed-beds, making dust-baths, and domestic pets are likely to disturb them, too. Wire-netting stretched over the rows is a good deterrent, if they have not had to be covered with cloches or polythene tunnels to protect against cold.

Sparrows can also tweak off the blossoms of climbing beans and will attack peapods, as will jays. A humming line is a help here, its high-pitched tone in wind discouraging them

Left In cool temperate climates, tomatoes are widely grown under glass, both commercially and by the home gardener.
Right Regular hoeing, especially when plants are young, will protect the vegetables against weeds.
Far right Plastic or wire netting is an inexpensive form of support for climbing plants like runner beans or peas.

without harm. Pigeons will ruthlessly attack winter crops of the cabbage family, completely destroying them, and a covering of plastic netting or a protective cage is the best answer. Blackbirds have a liking for tomatoes, and again protective netting is the best way of thwarting them.

Mice cause the most problems with pea-seeds and will eat a whole row in a night, so starting them in trays (flats) avoids this and overcomes the problem of patchy germination at the same time. In hard winters, they may also eat root vegetables left in the soil, especially if the plants' crowns are protected by straw or bracken.

Rabbits, like mice, will attack vegetables in winter, but regrettably they will also eat young plants, particularly lettuce, in spring and summer and if the vegetable garden abuts on to meadowland or woodland, precautions are essential. A wire-netting fence 120cm (4ft) high with another 30cm (1ft) length buried in the soil is the most efficient way of preventing their attacks; it must be securely staked and should have a mesh of about 25mm (1in) diameter.

Protection against cold
Cold protection is essential in cool temperate climates and in some warm ones where crops are being brought on early. For warming the soil in spring and for keeping the chill off tender plants usually until late in spring, there are a variety of cloches, made of clear and opaque glass or plastic. These are available in different sizes, but all with the same general shape of a barn or tent. Cloches should be easy to handle, light to carry, capable of being securely anchored if not heavy enough to resist wind, and capable of being stacked compactly when not in use. They should also be easy to put together. Polythene tunnels are a good substitute although the polythene discolours within two years and tends to get muddy at the sides easily. However, it costs little to replace with a new length.

Mobile cold protection of this kind is invaluable all through spring, in autumn starting early in that season and all through winter. Cloches and tunnels hasten summer crops, and a set of cloches can be in use right through the season with careful planning, bringing vegetables on much earlier and ensuring early setting of the fruiting vegetables. In spring, seeds and soil will be kept warm, and in autumn vegetables can be finished or protected from autumn rains until ready for cropping. Bracken or straw several centimetres or inches deep will stop the soil from freezing round root crops and a light layer over spinach, cauliflower, chicory and broad beans will keep off the worst frosts and snow.

Feeding vegetables
The feeding of vegetables is a subject which causes a great deal of argument. Some gardeners insist that the addition of a concentrated food supply in the form of fertilizers is essential; others maintain that the rotted organic matter added to the soil to improve its structure contains sufficient nutrient for the plants.

The food question is further complicated by the fact that every garden soil is unique and contains different quantities of minerals and has a different pH value; individual vegetables have been found to respond to variable quantities of different nutrients so feeding can become very precise indeed. In addition, there is a large and confusing range of proprietary fertilizers, all with different percentages of the three major plant foods: nitrogen (N), phosphorus (P) and potassium (K), as well as the trace elements. Some of these fertilizers act quickly, some, mostly the organic kind, last all season. Their method of application varies, too: some are given as dry dressings in granular mixtures, some are in powder form; some are mixed with water and applied to the soil, others are sprayed on to the leaves. Some are applied once, others twice in the season; some are given every week.

Types of fertilizers
However, as a general guideline, a compound fertilizer with equal quantities of N, P and K can be partly applied 10–14 days before planting or sowing, with the remainder of the dressing given halfway through the crop's growing season, watering in in both cases. The rate of application will be supplied on the container by the manufacturer, usually 112–140g per sq metre (4–5oz per sq yd) but some crops require much less, e.g. roots, and some much more, e.g. potatoes and the cabbage family.

If an organic fertilizer is used – for instance, blood, fish and bone – the whole dressing can be given at the start, since it is slow-acting. Where the soil is in good condition and has been regularly dressed with organic matter, extra feeding is unlikely to be necessary except in summer because of poor weather conditions – it may be extra nitrogen because of heavy rain, or extra potassium because of lack of sun.

Bonemeal is a popular vegetable-garden fertilizer in Britain and contains phosphorus and somewhat less nitrogen; it is slow acting and continues to be effective for about two years. It has an alkaline reaction and, if finely ground, is much more quickly available. Rates of application are about 56–112g per sq m (2–4oz per sq yd).

Dealing with problems

Vegetables can be subject to an enormous range of problems of various kinds, but in practice in most years, the worst offenders are few and are generally slugs and snails, aphids of various kinds, flea beetle, birds, mildew, and the water supply.

Provided crops are grown in good soil, given a strong start, and supplied with water in periods of low rainfall, their health should be good, and their natural immunity high – many modern varieties are bred for built-in resistance. Much can be done with good cultivation practices, such as not spacing too closely; sowing early or late as pests are usually around at standard sowing times; keeping the bed free of litter under which pests can hide; and using manual control wherever possible. Where fungal diseases appear, removal of affected parts as soon as they are seen can control them effectively. Weeds should be removed so that they do not supply cover and breeding grounds. If plants do become badly infested, dig up and destroy them.

Pests and diseases

If pesticides become necessary, be as sparing as possible and use those which have the shortest interval between application and harvest. Amongst these are: bioresmethrin, a synthetic analogue to the insecticide pyrethrum but more effective, for sucking pests, ants and caterpillars; benomyl and thiophanate-methyl, both systemic fungicides for mildew and grey mould (botrytis) control, and carbendazim or propicanizole for rust. The insecticides malathion (containing phosphorus) and derris need about two day's interval before crops can be harvested. For soil inhabiting insects, mostly caterpillars, trichlorphon is the least persistent. Between them, these few plus a specific for slugs (see below) should deal with any pest or fungus disease that appears; instructions for use will be on the container and should be strictly followed.

The general problems which occur throughout vegetable growing are as follows:

Aphids (greenfly, blackfly, mealy and root aphid). Tiny green, black or grey insects which cluster on tips of shoots and beneath young leaves, also older ones and flowers in bad attacks; feed by sucking sap from plants and cause yellowing, distortion and stunting, death if plant is small. Can be bad on young plants and during dry weather. Control by hand or spray bioresmethrin, derris or malathion. Root aphids feed on roots, plant grows slowly, becomes greyish green and may wilt; most common in dry, hot conditions; water with bioresmethrin or derris.

Caterpillars Mostly green, yellow, brown or multi-coloured kinds feed on leaves producing large holes; remove by hand, spray bioresmethrin or derris. Soil-inhabiting caterpillars such as leatherjackets (larvae of craneflies), chafer grubs and cutworms are grey-brown or white and feed on roots and crowns; white maggots may feed on bulbs or roots. Plants cease to grow, leaves become yellow, and plant wilts. Water other plants with trichlorphon as a precaution, allow two days

Left Magnesium deficiency in tomatoes causing premature withering of the leaves.
Right Leaf damage caused by caterpillars.
Far right Onions affected by grey mould caused by cool, damp or humid conditions.
Below Slug damage to radishes. Snails cause similar but less extensive damage.
Bottom Blackfly on a broad bean plant. These aphids feed on the sap of young shoots and so stunt the growth of the plant.

before harvesting.

Slugs and snails Eat large holes in all parts of plants, can destroy a row of seedlings overnight. They feed at night, so handpick after dusk or during daytime when they will be found hiding under stones, bricks, etc,. close to their source of food. Saucers of stale beer close to plants, gritty material in rings around plants or along rows about 10cm (4in) wide, traps of orange or grapefruit skins, hedgehogs and ducks, and proprietary herbal slug-destroyer are all deterrents (but harmless to warm-blooded creatures). Alternatively, use a chemical control containing methiocarb, but protect from hedgehogs, birds, cats and dogs.

Flea-beetle Tiny, hopping green-black beetle eats holes like shot-holes in the leaves of seedlings and young plants of the cabbage family, and can destroy a sowing overnight. Dust soil and plants with derris.

Red spider mite Tiny red, pink or transparent creatures suck sap from underside of leaves and stems, causing grey-green or yellowish colouring, withering and early fall. Plants become stunted, cease to grow, and webbing appears in bad attacks. Hand-lens needed to see them clearly. Worst in hot and dry conditions, usually on greenhouse plants. Spray malathion several times, as the makers instruct.

Whitefly Tiny white, moth-like creatures on under-surface of leaves; also appear as transparent discs, which are the young of the insect; sap sucking, they secrete a sticky residue on to leaves which becomes covered in sooty mould. Found on cabbages and on greenhouse vegetables. Spray bioresmethrin one or more times, and wash leaves.

Damping off A disease primarily of seedlings mostly found on seedlings in containers. The stems have a dark discolouration at soil-level, and the seedlings collapse and die; the disease infects from soil, and can be prevented by using sterilized soil; all proprietary seed composts should have been so treated. A watering with benomyl or Cheshunt compound will prevent further infection, though the affected seedlings will die.

Grey mould (*Botrytis cinerea*) A universal fungus disease, most prevalent in cool damp or humid conditions in which it spreads rapidly, characterized by grey fur which appears in spots and patches on leaves, stems, flowers and fruits; affected parts turn yellow, die and fall off or wither. Control by growing plants spaced more widely, removing affected parts as soon as seen, improving ventilation to dry out growing conditions and spraying with benomyl.

Powdery mildew White powdery patches characterize this fungus disease most likely to appear late in the growing season on leaves, and then stems, flowers and fruit. Dry soil, with damp or humid air, will encourage its appearance, and plant growth can be severely stunted or cease. At the end of summer many crops will have ceased to grow, and control is not essential; earlier, it should be removed when seen, and benomyl applied as for grey mould.

Soil-living fungus diseases There are many of these which can infect adult plants but unfortunately by the time it becomes apparent that a plant is infected, it is too late to treat it successfully. Such plants are invaded via their roots, and the fungus lives within the plant's tissues, blocking water uptake; wilting and yellowing leaves are common symptoms. Such plants should be dug up complete with roots and destroyed, and different crops should be grown on the site the following year.

Other problems include *bacterial soft rot*, mostly of roots, bulbs and tubers, when the vegetable is reduced to a liquid, offensively-smelling mass of decay; entry by the organism is gained through a previous injury from the soil and is often found in store. *Virus* diseases result in stunting, malformation and irregular discoloration, mostly yellow, and are transmitted largely by aphids, together with some other sap-sucking pests. There is no treatment and, for the sake of other plants, such individuals should be destroyed.

Brassica problems

The brassica family as a whole is subject to a good many pests and diseases. They include a variety of aphids, particularly mealy aphid, a species which is grey-blue in colour; whitefly; cabbage white butterfly caterpillars, coloured black, yellow, green and white, eat leaves, 3–4 broods between early summer and mid-autumn; cabbage roof-fly, white maggots eat roots of young plants, symptoms are wilting, blue-tinted leaves and slow growth of transplants, roots swollen and hollow, seen between mid-spring and midsummer, use soil insecticide at planting time or plastic discs on soil round plants; club-root, red or purple-tinted leaves wilt in hot weather, then permanently, roots swollen and distorted, no control – remove affected plants with roots and destroy, do not plant site with brassicas for at least 8 years, if possible replace soil in wide area; flea beetle, eats tiny round holes in seed-leaves and leaves of very young plants, can destroy them, beetle a tiny, black, hopping insect, dust derris round and on plants.

Special purposes

Rotating vegetable crops

Much has been written about the necessity for rotating vegetables, and this can sound complicated. Vegetables are often divided into categories for rotation purposes, but it is not essential to stick rigidly to these groups. Species can be swapped about between them, provided the cabbage and onion families are grown in different soil each year. The object is to avoid the onset of club-root disease of brassicas which is crippling and infests the soil for many years – there is no chemical control – and to prevent soil-borne fungus diseases of onions from building up.

However, one advantage of crop rotation is that all vegetables' roots penetrate to different depths and absorb different minerals – so that a constant change-round from year to year makes use of all that are available and prevents the depletion of one or more of them.

Three-year rotation plan

Vegetables can be divided into three, four or five groups, depending on how much ground is available and the number of types to be grown, but three is a convenient number for the average gardener and provides the basis for a three-year rotation plan. In the groups given as follows, Group 1 moves on to Group 2's site in the second year, Group 2 goes to Group 3's and so on, so that they go round and round. In this arrangement also, allowance is made for their manuring and feeding needs.

Group 1
Previous crop manured

Bean, French*
Beetroot*
Brussels sprouts*
Cabbage*
Calabrese*
Carrot†
Cauliflower*
Chicory
Onion†
Parsnip†
Pepper, sweet†
Potato†
Swede*
Sprouting broccoli*
Turnip*

Group 2
Manure given in winter

Artichoke, Jerusalem
Aubergine†
Celeriac
Endive
Fennel, Florence
Leek
Lettuce*
Radish
Sweetcorn
Spinach
Swiss chard
Tomato†

Group 3
Manure given in spring

Bean (not French)
Celery
Cucumber
Marrow
Pea
Pumpkin
Squash

* = lime in winter

† = add wood-ashes or sulphate of potash or potash-high fertilizer in spring

Both marrows and squashes need spring manuring.

Kohlrabi, which grows best in slight shade.

Vegetables for the 'hungry gap'

The time when fresh vegetables are least available is early spring to the end of late spring, with perhaps only spinach and the remains of the leeks and kale as the sources of green leaf, although sprouting broccoli is at its best. By sowing the right varieties, and with a little forethought as to sowing times, the selection can be greatly increased.

Asparagus (late spring)
Cabbage, spring
Kale
Leek (sown late)
Lettuce (sown early autumn & cloched) (late spring)

Onion, spring (late spring)
Radish (late spring)
Spinach, perpetual and winter
Sprouting broccoli
Turnip tops (sown early autumn)

Vegetables for shade

The majority of vegetables produce the best crops if grown in an open, sunny place, but there are a few which will succeed in slight shade, and some which can be prevented from bolting in the height of summer if sown where they will get shade as they approach maturity.

Artichoke, Jerusalem
Endive for a summer crop
Kohlrabi

Lettuce ⎫ if moist soil
Radish ⎭
Spinach, perpetual
Swiss chard

Quick-maturing crops

Vegetables which are only in the ground a short time can be put to good use by filling spaces temporarily empty between long standing vegetables or at the beginning of the season. Those which follow take two-three months to mature and are then mostly used up quickly; one or two are long standing.

Beetroot, for salads
Carrots
Chicory, red
Courgette
Endive
Kohlrabi

Lettuce, summer
Onion, spring
Radish
Spinach, perpetual and summer
Turnip

Scorzonera, a slow-maturing crop.

Maincrop turnips are slow maturing, early turnips quick maturing.

Slow-maturing crops

Some vegetables take a good deal of time before they are ready for harvesting, and in small vegetable gardens are not worth the space for the crop they produce, whether in size or length of harvesting. However, if they are particular favourites, it is usually possible to inter-crop them, that is, sow one of the quick-growing vegetables in between the main crop (see below). The ones given below take six months or more to mature.

Artichokes, Jerusalem	Onion, Japanese
Broad beans, autumn-sown	Parsnip
Brussels sprouts	Potato, maincrop
Cabbage	Salsify
Cauliflower	Scorzonera
Celeriac	Sprouting broccoli
Celery, winter	Swede
Kale	Turnip, maincrop

Summer salads

All the vegetables which follow can be eaten raw and are best if eaten immediately after they are gathered and prepared. The minimum of washing should be done to avoid loss of vitamins and minerals, and they should be torn, not cut.

Chinese cabbage	Onion
Carrot	Pea, mange-tout
Celery, self-blanching	Pepper, sweet
Cucumber	Radish
Chicory	Tomato
Lettuce	

Winter salads

The number of vegetables that can be used raw for winter salads is the same as for summer; there is no need to be short of fresh material at this time. Salads, of course, can also be made from cooked vegetables and eaten cold at whatever the time of the year. Beetroot, French beans, asparagus, fennel, peas and sweetcorn are examples.

Cabbage,white	Endive
Chinese cabbage	Lettuce
Carrot	Onion
Celeriac	Radish, winter
Celery, winter	Sprouting broccoli
Chicory, red, and chicons	

Vegetables for intercropping

The crops which take a long time to mature may not be worth growing in small areas, but their expensive use of space can be overcome to some extent by sowing or planting a quickly-growing vegetable between the young plants while they are small, or between rows. Plant or thin to a little further apart, and remember that both crops must have the light, air and room that they need. The following are examples and there are many more combinations which can be worked out according to individual needs.

Sprouting broccoli with lettuce	Lettuce with sweetcorn
Parsnips with early radish	French beans with winter cabbage
Early dwarf pea with marrow	Broad beans with cauliflower
Fennel with leek	Runner beans with forcing chicory

Leafy winter vegetables

Vegetables which will provide fresh green material in winter are very valuable as they contain a good deal of iron and other minerals, and the vitamins C and A to improve resistance to infections and keep the skin in good condition in low temperatures. The B complex of vitamins help perform many important functions in the body, ensuring our efficient nervous system, good digestion, stamina and the maintenance of tissues. It is another vitamin particularly found in this type of crop.

Brussels sprouts	Lettuce
Cabbage	Swiss chard
Cauliflower	Spinach, winter and perpetual
Chicory, Italian	Sprouting broccoli
Kale	Turnip tops

Artichoke, Globe

(Cynara scolymus) COMPOSITAE

Globe artichokes are the kind whose flowerheads are eaten, unlike the Jerusalem variety, of which the tubers are the edible part. A globe artichoke plant is a large one, spreading to at least 90cm (3ft), and being about 90–120cm (3–4ft) high when in flower and is sufficiently ornamental to be allowed into the herbaceous border, where it will flower from midsummer until autumn.

The part eaten is the immature flowerhead, cut before the flowers appear, while it still consists only of overlapping, green, fleshy scales. The scales are pulled off singly, after cooking in boiling water for about 40 minutes until tender, and dipped in sauce or melted butter, then the base of the scale is drawn through the teeth. The tender base of each head – the heart – is also eaten, but not the 'choke' in the centre, which is inedible.

Globe artichokes are slightly tender, and are found growing throughout the Mediterranean region; they have never been found wild, and are probably a cultivated form of the cardoon, *C. cardunculus*, which is a native of the area. Globe artichokes have always been considered a delicacy; they have been widely grown in southern Europe for many centuries.

Culinary uses

While boiling them is the accepted method of cooking, they can in fact be stuffed, fried, baked, used in salads and eaten hot or cold. To prepare them, most of the stalk should be cut off, and the points of the leaves or scales trimmed back with scissors; remove the inedible choke (with tiny young artichokes this is not necessary), rub leaves and heart with lemon and drop into cold water until ready to cook. They contain small quantities of minerals, have a low calorie value and also contain some vitamins, chiefly vitamin A.

Cultivation

Globe artichokes are perennial herbaceous plants, and will supply heads for several years. Their main needs are for deep fertile soils and plenty of moisture, together with protection from cold of the moderate to severe grade. Cold, badly drained ground or frost pockets are not suitable sites for them, and they are difficult to grow outdoors in Scandinavia and northern Britain. However, once established, they need little attention.

CULTIVATION

Site and soil sunny, sheltered from wind, warm; moist, deep, fertile, well-drained soil

Soil preparation dig the soil deeply to two spades' depth, fork up the base of the hole and add well-rotted organic matter – farmyard manure, garden compost or composted seaweed are excellent; then return the soil. Fork a dressing of bonemeal into it during late winter–early spring

Plant put in plants in mid spring 120cm (4ft) apart, and 90–120cm (3–4ft) between rows; plant firmly 10cm (4in) deep, with the growing tips just above soil level

Summer care keep free of weeds; watch for snails and slugs eating the young shoots; mulch on to moist soil after planting, and thereafter in mid spring each year; keep well watered in dry weather; in early summer remove all shoots from the base except the five strongest

Winter care remove largest leaves, earth up the inner ones, and cover with bracken or straw when frost is forecast, remove this covering in mild spells. Remove earthed up soil in early spring

Harvest take only 3 or 4 heads in first season, remove all the other flowering stems as soon as seen; thereafter cut from early summer–autumn. Use large central head, remove side-buds when about 5cm (2in) long – eat raw or pickle them. Cut main heads with 15cm (6in) stalk, when lowest scales are just beginning to open; rest of head should still be tightly closed

Troubles petal blight, grey mould, *Botrytis cinerea*, pale brown spots on petals and scales, whole flower-head rots; spray propiconozole; earwigs; greenfly, slugs and snails

Increase use shoots left on the plant for increase; they are suckers, and are detached with roots and 'heel' of old stock in mid autumn or mid spring. Replant firmly at original depth

ARTICHOKE, JERUSALEM
(Helianthus tuberosus) COMPOSITAE

They are natives of North America, being first found in Nova Scotia by a group of French explorers, and were introduced to Europe, including Britain, at the beginning of the 17th century. The common name Jerusalem is thought to be derived from the Italian *girasole*, meaning sunflower, to which they are closely related. The Victorians called the soup made from them 'Palestine soup'.

Like the sunflower (*Helianthus annuus*) they grow tall, to 2.4m (8ft) if allowed to and, with their large leaves, can make a useful windbreak 60 or 75cm (2 or 2½ft) wide. They rarely flower in northern Europe, and when they do, the flower is a small yellow one; hot dry summers are needed to induce blooming. The tubers form late in the season, and the first digging should never be made until the stems have virtually finished elongating.

Nutritional value
Jerusalem artichokes are a welcome variation on the winter root crops, particularly as they contain vitamin C and many mineral salts, some protein, and a good deal of inulin, but have a low calorie count.

Culinary uses
The tuber, which looks like a nobbly potato, is the part eaten and makes an excellent and tasty soup. Jerusalem artichokes can also be boiled, baked or steamed, adding a little lemon juice to them to prevent discoloration, or served raw, grated or cut into julienne strips, in a vinaigrette dressing.

Cultivation
Cultivation is simplicity itself; in severely cold winters the tubers take a couple of months to appear above ground level if planted at the normal time of late winter, but otherwise the main point to watch is that all the tubers including the tiny ones, are dug up in winter; if not, they infest the ground in future years and are difficult to eradicate.

CULTIVATION

Site and soil sun or shade; any soil provided it is not waterlogged, though good soils will produce bigger tubers. In windy positions plants will need staking

Soil preparation dig in organic matter in late autumn; add a dressing of wood-ashes a few days before planting

Plant plant egg-sized tubers in late winter–mid spring, 10–15cm (4–6in) deep, 60cm (2ft) apart, with 90cm (3ft) between rows. Expect sprouting 2–4 weeks later in average weather

Summer care watch for slugs/snail damage on young emerging shoots; control weeds; stake plants in midsummer if in windy positions; remove flowers if they appear

Winter care at the beginning of winter, cut off stems close to ground level; cover the soil to prevent frost making it iron-hard and therefore impossible to dig

Harvest dig as required from late autumn–early spring; store in dry ashes or sand for up to 2 months, but flavour will not be as good. Use the largest of the last-dug for the new season's plants

Troubles slugs/snails on young shoots

ESSENTIAL FACTS

Type of plant annual in cool temperate climates.
Part eaten tuber.
In season late autumn–early spring.
Yield 2–2½kg (4–5lb) per plant; about 8 tubers/½kg (1lb)
Time from sowing to harvest 8–9 months
Size plants 1.8–2.4m (6–8ft) tall, 60–75cm (2–2½ft) wide; tubers 5–7 × 2.5–4cm (2–3 × 1–1½in)
Hardiness completely hardy

RECIPE
Exotic

Artichokes with scallops
6 scallops, whites cut into 2 or 3 rounds
250g/½lb Jerusalem artichokes
150ml/¼ pint milk
150ml/¼ pint thick cream
25g/1oz butter
25g/1oz flour
salt and freshly ground black pepper
3–4tbls crushed, crisply fried bacon or fresh breadcrumbs fried in garlic butter

Cut the artichokes into thickish slices and simmer in milk and cream until just tender. Remove onto scallop shells (or scallop-shaped dishes), cover and keep warm. Gently poach the scallop rounds and coral (left whole) in the milk and cream liquid for about 5–7 minutes or until just cooked. Remove, add to the artichokes and keep warm. Make a roux with butter and flour, add cream and milk liquid and simmer until sauce has thickened somewhat. Add salt and pepper to taste, pour sauce over scallops and artichokes and sprinkle with hot crushed bacon or breadcrumbs.

Asparagus

(Asparagus officinalis) LILIACEAE

The most delectable of vegetables, asparagus has the best flavour when eaten fresh. This, plus the fact that asparagus is expensive to buy, makes it an obligatory inhabitant of the vegetable garden. It is a long-living perennial plant and the choice of site needs extra care; 20 years is the minimum life in a garden, with the best crops as it matures. With care, the shoots can be gathered for 2–3 months, though the average season is about 6–8 weeks.

The Romans and Greeks cultivated this coastal plant over 2000 years ago, and it can still be found growing wild in Britain and Europe, and as far east as central Asia. A member of the lily family, it was being grown in English gardens in 1534.

Nutritional value

The specific name of asparagus gives it away; it was once an official medicinal herb, noted for its diuretic and laxative properties. Its highest mineral content is potassium, and it contains appreciable quantities of vitamin C and carotene.

Culinary uses

The green tips are the tenderest and best-flavoured part – the white stem becomes progressively tougher, though if 15mm (½in) wide they will be more succulent than the pencil-thin stems, often long and green and used for soup. Generally, asparagus is boiled or steamed and served hot or cold with melted butter, mayonnaise or other types of sauces.

Cultivation

The soil for asparagus needs to be thoroughly prepared with trenching, manuring and complete clearance of perennial weeds. Once the plants are in, removal of such weeds as bindweed, couchgrass or ground-elder is virtually impossible, and the asparagus will never achieve its cropping potential.

Too much cutting of the stems in one season results in the production of many small thin shoots the following year; cutting should stop between the middle and end of early summer, the remaining spears being allowed to grow full length into fern, so that the crown and roots can be fed and developed. Although hardy, in really cold winters, a protective straw mulch is advisable, and in late spring, frosts can occasionally damage the tips. Male plants give the heaviest yield.

The only serious problem may be asparagus beetle, which feeds sufficiently voraciously to weaken the plants seriously.

CULTIVATION

Site and soil open, sunny, sheltered from wind, not a frost-pocket; well-drained, sandy, deep, fertile

Soil preparation dig 2 spades deep in autumn, mix rotted organic matter with returned soil; a 180cm (6ft) wide bed will take 2 rows 90cm (36in) apart of staggered plants, with 45cm (1½ft) between plants in the rows. Add lime later, in winter, if soil acid; fork in general fertilizer in early spring. Clear out all weeds rigorously. If drainage bad, add old mortar rubble or sand, while digging

Plant use 1– 2 year-old plants, and plant in mid-spring, late spring if season wet and cold. Keep the plant roots moist in plastic bags until actually planting. Dig out trench 20cm (8in) deep, 30cm (12in) wide, and return some soil to form mound down centre of trench, 7cm (3in) high at highest point. Put plants on this 45cm (1½ft) apart, with roots spread out well down sides of mound. Allow 90cm (3ft) between trenches. Fill in crumbled soil over plants 5cm (2in) deep, and firm, water well; add more gradually through season; do not allow to dry out at any time

Summer care hoe and weed regularly; keep well watered in drought; allow all shoots to grow in first year and support with canes; in second summer, allow most shoots to grow; flatten ridges in midsummer

Winter care cut fern down when it turns yellow to about 25mm (1in); clean up bed and apply organic matter; draw low ridge of soil over plants; dress with woodash late winter–early spring, and nitrogen in early spring

Harvest in 1st summer of planting, do not cut any shoots; in 2nd spring– summer cut only three or four from each 3-year-old plant none from the 2-year-old; 3rd season, cut few more from by then 4-year-old plants, and in the 4th season, cut as required. Cut shoots about 15cm (6in) long when tips just above ridge level, to give mostly white spears; cut when spear about 10cm (4in) above for white and green ones, and when 20cm (8in) above for green ones. Cut carefully to avoid cutting other shoots not above ground. Cut all suitable shoots, every day if necessary, and store in refrigerator, otherwise production ceases; complete harvesting by end of early summer

Troubles Asparagus beetle, orange-red and black, 6mm (¼in) long, grey, hump-backed larvae, appear early summer, eat shoots and foliage; hand-pick; spray derris several times; slugs/snails eat young shoots; red-brown raised spots on leaves, cut shoots to ground level and destroy; frost in spring

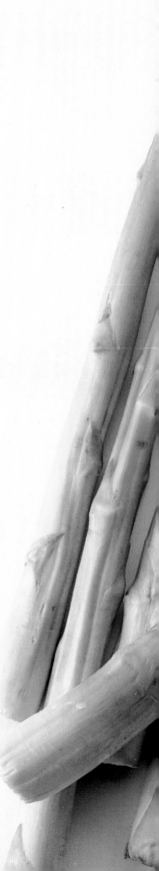

Type of plant perennial
Part eaten tip of stem
In season mid-spring–early
summer
Yield 15–20 spears per plant
**Time from planting 2-year-old
crowns to harvest** 12–14 months
Size plants, 120–150cm (4–5ft) tall
and 60–90cm (2–3ft) wide
Hardiness tips of young shoots
tender

RECIPE
Exotic

Asparagus with mustard sauce
1kg/2lbs asparagus
2oz/50g butter, softened
4 egg yolks
*2tsp coarse-grained French
     mustard*
Peel the lower part of each
asparagus, trim the bases and tie
into four loose bundles. Stand
upright in a tall saucepan, add
boiling water and salt but leave the
tips uncovered. Cover with a lid (or
domed foil) and boil for about 20
minutes or until the stalk is tender
when pierced with a sharp knife.
Drain. Remove to warmed serving
dish and keep warm while you
make the sauce. Whisk the egg
yolks until light and fluffy, add the
softened butter and mustard and
whisk again until well blended.
Serve immediately over the
asparagus.

Aubergine

(Eggplant, Brinjal, Jew's-apple, *Solanum melongena* 'Esculentum') SOLANACEAE

Aubergines are annuals, natives of South Africa and Southern Asia, and belong to the same family as the potato and tomato. They were described in Gerard's *Herbal* of 1597 where the white egg-shaped kind was called Raging or Madde Apples (*Mala insana*), and reached America in about 1850. They have long been popular in India, the Middle East and southern Europe but only recently in Britain and northern Europe, where they are now widely available. Even so, they seem to be a vegetable that you either love or hate. Indeed, Gerard advised his readers that: 'it is therefore better to esteem this plant and have it in your garden for your pleasure and the rareness thereof than for any vertue or good qualities yet knowne.'

Varieties

The aubergine varieties commonly grown in cool temperate climates have an upright habit of growth and long, sausage-shaped or long-oval shaped, purple fruits, but there are also varieties with white fruit shaped exactly like a hen's egg, though slightly larger. In fact there are many different kinds grown in tropical climates, whose fruit varies in shape between round and long, and which may be striped, white, black and varying shades of purple. Growth habit can also vary between sprawling, dwarf and branching, and tall; the species can grow to 2.4 metres (8ft) in its native habitat.

Nutritional value

However, they are completely edible, and contain a variety of mineral salts, vitamin C and dietary fibre, and have low calorie and carbohydrate values.

Culinary uses

Aubergines are widely used in stews and often served cold, having been first cooked in oil with herbs and spices. They are also frequenty stuffed with savoury mixtures and added to curries. They can add a rich texture and flavour to many dishes. The somewhat bitter taste of the raw vegetable can be eliminated to some extent by 'sweating' them with salt sprinkled on to the sliced surfaces, wiping off the liquid which results before cooking them.

Cultivation

To cultivate them successfully, in cool temperate climates, aubergines require the same degree of protection given to tomatoes. They are difficult to mature out of doors, unlike tomatoes, unless the summer is exceptionally good, and are best grown in a greenhouse, after starting the seeds off with artificial warmth in early spring. The plants have large, soft leaves, sparsely prickly stems and fruit stalks, and attractive violet or white flowers like those of the tomato, but larger. The plants need a long growing season to mature and, if sown in early spring, will not crop until late summer.

CULTIVATION

Site and soil with cold greenhouse protection, or in sunny sheltered position outdoors; moist, fertile, well-drained soil

Soil preparation in containers, use J.I. No 3 for final potting; dig soil 1 spade deep and fork in rotted organic matter in winter; in spring mix in general fertilizer 2 weeks before planting

Sow seed late winter–early spring in 18°C (65°F) thinly in trays, or singly in 5cm (2in) pots, sow 6mm (¼in) deep

Prick out into 7.5cm (3in) pots of J.I. No 1, and then into 12cm (5in), finally into 23cm (9in) pots; outdoors or in greenhouse soil, plant when 5cm (2in) pots outgrown. Space 60cm (2ft) apart each way; plant out when no risk of frost, and mulch immediately. Water in after each move

Care watch for snails/slugs when just planted out or potted; keep well watered in dry weather; liquid-feed in containers as first fruit begins to swell

Train/prune stake stems; remove growing point at 15–30cm (6–12in) tall if sideshoots wanted

Harvest when fruit have finished swelling/elongating and still shiny; cut fruit stalk with scissors; purple varieties are purple from the time of setting

Troubles grey mould in cool dull, or wet conditions, on flower end of fruit and calyx; red spider mite; whitefly; greenfly

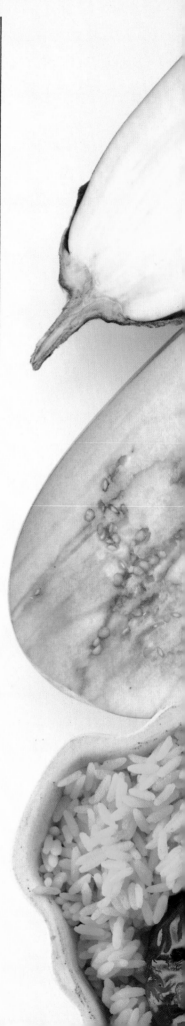

Type of plant perennial grown as annual
Part eaten fruit
In season late summer–mid autumn
Yield approx. 4 fruits/plant
Time from sowing to harvest about 16–18 weeks
Size plants 90 × 60cm (3 × 2ft) fruit 7–15cm (3–6in) long
Hardiness tender
Seed viable 6 years
Germination period 10–20 days

RECIPE
Exotic

Imam Bayildi

The Turkish title of this recipe means 'The Imam (holy man) fainted' allegedly with pleasure after simply inhaling this dish's fragrance.

4 aubergines
4 tomatoes, skinned and chopped
4 medium-sized onions, finely chopped
50g/2oz currants
2 garlic cloves, chopped finely
3-4 slices sweet red pepper
1 bay leaf
1tbls chopped parsley
salt and black pepper
½tsp mixed spices
olive oil
30–40g/1–1½oz toasted pine nuts

Make 4–5 lengthways cuts partially through each aubergine, salt, and leave to drain upside down. Put currants to soak in water. Fry onions and garlic in oil until they change colour, add tomatoes, mixed spice, parsley, salt and pepper, simmer until soft. Add currants, cook 2–3 minutes until oil is absorbed and mixture fairly dry. Cool, then stuff aubergines with mixture through slits. Top each slit with a thin slice of red pepper, if desired. Put into ovenproof dish, and pour olive oil over them to come a little way up aubergines. Add bayleaf, cover, and cook slowly until aubergine are soft, and slightly sticky residue at bottom of dish. Leave to cool in oil. To serve, drain off the oil and sprinkle each aubergine with pine nuts and additional parsley. Serve on a bed of saffron rice, if desired.

Beans

BROAD BEAN
(Fava bean, English bean, *Vicia faba*) LEGUMINOSAE

Broad beans are one of the earliest vegetables to mature in cool temperate climates, and one of the easiest and least complicated to grow. They are leafy plants with several stems, varying in height from 30–120cm (1–4ft) depending on variety, and only need the minimum of support. Their white, black-blotched flowers are heavily fragrant and much visited by bees during early-midsummer.

The history of this vegetable is recorded as far back as the ancient Egyptians; the Romans knew and valued it, and it was certainly grown in Europe during the 15th century, and probably earlier.

Nutritional value
The likely reason for its long history is its ease of cultivation combined with its high energy value, high carotene content and high dietary fibre (where the pods are also eaten), though its food value was probably not described in these terms by the ancients.

Culinary uses
The young beans can be picked when the pods are 5–7cm (2–3in) long and the tender pods eaten whole; later the large pale green, roundish beans are shelled and cooked without the pods, and finally the beans can be dried, when they will be beige-coloured, stored, and cooked as required in winter (these are commonly used in Egyptian and Middle Eastern cooking where they are known as *ful medames*). Use the top leaves of the plants like spinach; cook beans and the young pods by boiling or steaming.

Cultivation
Broad beans are annuals, and can be sown outdoors in spring, or late autumn to overwinter, but the latter is risky as severe cold or waterlogging can kill them, and cloche protection may make them come on too quickly to resist late spring frosts, and flower before pollinating insects have emerged. Since the crop from these plants will only be two–three weeks earlier than the spring-sown one, the advantage is not great, but gardeners in sheltered mild districts could well try it. Pods are produced in clusters all the way up the stems, and point upwards; they do not hang down, as French and runner beans do, and are well concealed in the foliage.

CULTIVATION

Site and soil sun/little shade; moist soils, not waterlogged in winter
Soil preparation dig 1 spade deep, add organic matter in early spring
Sow outdoors in spring, in late autumn in mild gardens; space 20–25 × 38–45cm (8–10 × 15–18in) apart, depending on variety, sow 5cm (2in) deep; sow extra seed at end of row, as gaps likely
Care gap up plants; keep down weeds; water if dry when beans swelling
Prune/train support taller varieties with canes, or use wire both sides of the plants the length of the row attached to posts, when plants 60cm (2ft) tall; remove basal sideshoots; nip out tips of stems as beans start to swell
Harvest when pods about 7cm (3in) long, cook whole; when pods full length and scar on bean is still white or green, cook beans (seeds) only; dry for winter use
Troubles blackfly on tips of plants; nip out tips in advance as above; chocolate spot, on leaves and pods, destroy plants badly affected, spray rest with propicanozole; both troubles can ruin crop

ESSENTIAL FACTS

Type of plant annual
Part eaten beans, young pods
In season early–late summer
Yield 2.5kg (5lb)/3m (10ft) row
Time from sowing to harvest 3–4 months spring-sown; 6½ months autumn-sown
Size plants 30–120cm (1–4ft) tall, 20–30cm (8–12in) wide, pods up to 23cm (9in) long
Hardiness hardy
Seed viable 2 years
Germination period 10–20 days

RECIPE
Economical

Broad beans and rice
This popular Middle Eastern dish can be served on its own with a bowl of yoghurt and a green salad or as an accompaniment to chicken or lamb.
2 large onions, chopped
2–3 cloves garlic, finely chopped
3tbls oil
1kg/2lb broad beans, shelled
1tbls ground coriander seeds
1tsp turmeric or powdered saffron
250g/½lb long grain rice
½ litre/¾ pint water
salt and pepper
fresh coriander leaves, chopped
Heat the oil in a large saucepan, add the onions and garlic and cook until soft. Add the coriander and turmeric and cook for another minute or so, stirring, before adding the beans. Stir well for another minute or two before stirring in the rice, salt and pepper and water. Bring to the boil, cover and simmer gently for about 20 minutes without lifting the lid. At the end of the cooking time, the rice should be tender and the liquid absorbed. Turn off the heat and leave for about 5–10 minutes before serving, garnished with coriander leaves.

Imperial green longpod

FRENCH BEAN
(Snap beans, string bean, kidney bean, shell bean, haricot vert *Phaseolus vulgaris*) LEGUMINOSAE

It is likely that French beans were grown by the Incas, and later the American Indians, reaching Britain by way of the Continent during Elizabeth I's reign, probably introduced by Flemish and French immigrants.

The French bean is actually a native of South America, but it is the most popular bean in France, where it is eaten all through the summer, unlike the runner (pole) beans which hold first place in Britain and America. The plant is mostly bushy and small, and it cannot stand frost.

Varieties
The pods are long and narrow, round in section, but even so are much smaller than runners, with a more delicate texture and flavour, and are eaten whole while still juicy and tender. When mature, the beans inside can be eaten, like peas, and are called flageolets, or French horticultural beans, and if these are left until the pods turn dry and yellow, they can be dried and are then haricot beans. There are special varieties for producing haricot beans. The pods may also be coloured: yellow (the waxpod types), purple, and speckled purple as well as green, and the beans themselves can be white, brown, black, or speckled and blotched brown or light brown or white.

Nutritional value
They are a good source of vitamin C and carotene, contain a variety of mineral salts, some dietary fibre, but very little protein unless eaten as haricots, when the level is raised more than six times.

Culinary uses
Boiling or steaming is the accepted method of cooking, although they are frequently stir-fried in Oriental cooking. Served cold they make an excellent basis for a salad.

Cultivation
Cultivation points to note are that slugs and snails will eat the pods if they lie along the ground. Plenty of moisture at all times is important – it ensures a good set of flowers, it prolongs the cropping season which is otherwise rather short, and if watering is kept up, after all the beans have been picked, a second smaller crop will ensue, provided fertilizer is added to the water.

CULTIVATION
Site and soil sun, shelter; moist soils, not too heavy or acid, use site manured for previous crop
Soil preparation dig 1 spade deep in autumn, add bonemeal in late winter; use lime some weeks earlier if soil acid; warm the soil with cloches before sowing
Sow early–mid-spring under cloches in warm gardens, or unprotected, late spring–midsummer; sow seed 5cm (2in) deep, at 10cm (4in) apart, rows 45cm (1½ft) apart
Care protect from slugs/snails as shoots come through soil; keep well watered throughout; mulch in early summer; remove weeds
Prune/train support plants as they grow, the bushy kinds with short brushwood, the climbing ones with canes, string or netting
Harvest when pods 10cm (4in) long, and still easy to snap in half; pick frequently for about 5 weeks; cut pods off with scissors as plant easily pulled out of ground
Troubles slugs/snails; blackfly

ESSENTIAL FACTS
Type of plant half-hardy annual
Part eaten pod, bean-seed
In season maincrop, midsummer–mid autumn
Yield 3–4kg (6–8lb)/3m (10ft)
Size bush-plants 23 × 23cm (9 × 9in); climbing 150 × 23cm (60 × 9in); pods 10–15cm (4–6in)
Hardiness tender
Seed viable 2 years
Germination period 10–14 days

RECIPE
Healthy
Mix boiled and cooled French beans with tomatoes and raw red sweet peppers; cut the peppers and tomatoes into strips. Put all in a vinaigrette containing garlic for a few minutes, then mound up on a serving dish, surround the base with sliced hard-boiled eggs and garnish with parsley.

Exotic
Spiced French beans with chilli
1lb/500g French beans, topped and tailed
1 clove garlic, minced
small piece of ginger, peeled and finely chopped
1–2 chilli, seeded and finely chopped
2tbls vegetable oil
1tbls light soy sauce
1tbls sesame oil
Blanch the beans in boiling water for 1–2 minutes. Drain and run under cold tap to prevent beans from cooking further. Heat a wok or saucepan over a high flame, reduce heat and add vegetable oil. Add garlic, ginger and chillis, cook for 1 to 2 minutes, stirring so they do not burn. Add the beans and soy sauce and stir fry for a further 3 or 4 minutes until the beans are cooked but still crunchy. Remove from the wok or pan with a slotted spoon into serving dish. Sprinkle with sesame oil and serve immediately.

Tendergreen

The Prince

CULTIVATION

Site and soil sun/little shade; sheltered from wind; most soils, preferably deep, moist and fertile

Soil preparation dig 2 spades deep in spring, mix organic matter with the soil as it is returned; fork in potash-high fertilizer about 3 weeks before sowing

Sow in individual 7cm (3in) pots in mid-spring and protect with frame or greenhouse; sow outdoors in late spring, and protect with cloches; space 23cm (9in) apart, 45cm (18in) between rows; sow 5cm (2in) deep

Transplant plant out in late spring and protect with cloches

Care watch for slugs/snails as germination occurs; remove weeds; mulch in early summer; water heavily, twice a week in hot dry weather, especially if this occurs when flower buds showing

Prune/train supply supports when sowing/planting, e.g. canes, wigwams, netting, vertical strings from wires attached to posts; train plants round supports manually to start them off; pinch out growing tips at 1.8m–2m (6–7ft)

Harvest pick beans when about 17cm (7in) long, use scissors; keep cutting every 2–3 days to maintain production; deep-freeze or salt surplus

Troubles flowers not setting, lack of water at roots, strong or cold wind discouraging pollinating insects; flowers taken by sparrows; slugs/snails on seedlings; blackfly, greenfly

BEANS, RUNNER
(Pole bean, stick bean *Phaseolus coccineus*) LEGUMINOSAE

In their native country of South America, runner beans are perennial plants, but in cool temperate climates they are not hardy and so are grown as annuals, from seed (beans) grown afresh every spring. They produce a thick, almost tuberous root, which can be dug up in autumn and stored like dahlias for replanting in the spring, but growing from seed each year is much less trouble.

They climb vigorously by twining to a height of 3m (10ft) if allowed to, but in gardens they crop better and are more convenient to deal with if restricted to about 2m (6ft). Since they climb so quickly, they are a good plant to provide temporary cover for walls or fences in the ornamental garden, especially as they have masses of red, white or salmon-coloured flowers from midsummer and green or purple pods until mid-autumn. In fact, although introduced to Britain in 1633, they were not regarded as edible until about a hundred years later, and were in the meanwhile grown purely for decorative purposes.

Runner beans crop for over three months during the summer, making them one of the most popular vegetables; they are particularly good to eat, especially if picked before they become extremely long. Those available in shops have usually been allowed to grow too big, when they are tough, stringy and have lost much of their flavour. Picked when less than a foot in length, they are at their best-flavoured and most tender – frequent picking, moreover, encourages the production of more beans.

Nutritional value
Nutritionally they supply a good deal of carotene and vitamin C, some protein and dietary fibre, little in the way of calories or fat, and a variety of mineral salts. If eaten raw, as they can be when about 10cm (4in) long, their food value is even better.

Culinary uses
Boiling or steaming is the accepted method of cooking, but they can be eaten cold as well as hot, and in salads – the raw pods eaten whole when young are also a good salad ingredient.

Cultivation
Runner beans do not present any difficulty in cultivation, provided they are kept free of frost, and provided the soil where they are to grow is moisture-retentive and has plenty of water in reserve deep down, otherwise the flowers will drop without setting, unless heavily and frequently watered in dry, hot weather.

ESSENTIAL FACTS
Type of plant perennial grown as annual
Part eaten pods and immature beans
In season midsummer–mid-autumn
Yield 0.75–1kg (1½–2lb) per plant
Time from sowing to harvest 12–14 weeks
Size plant 1.8–3m (6–10ft); pods 15–45cm (6–18in)
Hardiness half-hardy
Seed viable 2 years
Germination period 10–15 days

RECIPE
Healthy

Soup au pistou
This is really a mixed vegetable soup. The traditional version contains at least two types of beans and the essential ingredient, fresh basil.
125g/4 oz young runner beans, sliced
125g/4 oz courgettes, diced
250g/8oz tomatoes, peeled
125g/4oz Spanish onions
125g/4oz carrots, peeled and sliced
125g/4oz young turnips, peeled and diced
200g/6oz haricot beans, cooked
125g/4oz vermicelli
125g/4oz celery, sliced
2–3 sprigs basil, chopped
Pistou
100g/3oz fresh basil leaves
8tbls olive oil
25g/1oz pine nuts
2 cloves garlic, crushed
75g/2½oz grated Parmesan cheese
In a large saucepan, cover all the vegetables, except vermicelli and haricot beans, with water. Add basil and seasoning and simmer for about 30–40 minutes. Add cooked haricot beans and pasta and cook for a further 10–15 minutes until pasta is tender. Meanwhile, make the pistou in a blender or mortar. Grind the basil, pine nuts and garlic to a paste, add the olive oil gradually, then the cheese. Mix the pistou into the soup tureen or serve separately with the soup. Serve extra Parmesan cheese, if desired.

DRIED BEANS

Shell beans are dried bean seeds, whereas pod beans are eaten in the form of fresh bean seeds or the pods themselves are eaten with the immature seed inside. Shell beans are a highly concentrated form of protein, and are also low in fat and high in dietary fibre. They are an essential part of a vegetarian diet since they can provide much of the protein and other nutrients supplied by meat. They have a long shelf life, if kept dry, and have endless uses, especially if combined with spices and herbs, or with some of the more unusual vegetables. Dried beans form the basis of many well-known dishes such as cassoulet, Boston baked beans and chilli con carne to name but a few. All dried beans need to be soaked overnight before cooking and cooking time varies according to the type of bean and its shelf age. Beans which have been stored for longer than six months will rarely soften, no matter how prolonged the cooking. Always add salt at the end of the cooking time to prevent the beans from hardening.

BLACK-EYE BEAN
(Cowpea, *Vigna unguiculata. V. sinensis, Dolichos unguiculatus*, LEGUMINOSAE)

The variety of common and botanical names indicate the worldwide distribution and popularity of this small bean, which is cream-coloured with a pronounced black spot, about the size of a haricot bean. It has a strong flavour, and should therefore be used with care in soups and casseroles; it makes a good ingredient if used cold for mixed bean or other salads, and can be used hot as a side vegetable. Soak overnight, and cook for about 40–45 minutes.

HARICOT BEAN
(*Phaseolus vulgaris* variety, LEGUMINOSAE)

Haricot beans are the dried beans taken from the pods of French beans which have matured and turned yellow. They are small and oval and a creamy-white in colour. They are a common ingredient in a wide variety of French dishes and are the traditional accompaniment to roast lamb in many parts of France. In the US, they are known as navy beans. After soaking, they should be cooked for about 1 hour.

FLAGEOLETS
(French horticultural beans, green beans, *Phaseolus vulgaris*, LEGUMINOSAE)

These are the beans found inside pods which are only just fully grown, and they are harvested while the bean seeds are still pale green. Size is not quite 1.5cm (½in) long, and they have a particularly delicate flavour as they are gathered while still young. Cooking time is slightly longer than most, about 1–1¼ hours.

RED KIDNEY BEAN
(*Phaseolus vulgaris* variety, LEGUMINOSAE)

The familiar red bean of chilli con carne really needs no description beyond saying that they are about 1.5cm (½in) long and are red to brown-red in colour. An ingredient of many North American dishes, they blend well with meat casseroles, in which they can replace some of the meat and thus provide a cheaper but just as nutritious a dish; they can also be used in salads of mixed beans and provide an attractive variation in colour. Cooking time should be about 45 minutes.

BORLOTTO
(French or kidney bean variety, *Phaseolus vulgaris*, LEGUMINOSAE)

Another popular shell bean, this is comparatively large, being about 2cm (¾in) long. It might be striped red on a whitish background, or plain red. After the standard soaking, it needs cooking for at least 1 hour.

CANNELLINO
(*Phaseolus vulgaris* variety, LEGUMINOSAE)

Smaller than those previously mentioned, these beans are rather long and narrow in shape, and develop in a narrow pointed pod. The colour is universally creamy white, and cooking time after soaking is ¾–1 hour.

Runner beans

Dutch brown beans

Lima beans

Red kidney beans

Black-eye beans

Borlotto beans

Cannellino beans

Flageolets

Beetroot

(Beta vulgaris 'Crassa'*)* CHENOPODIACEAE

The beetroot is a curious vegetable, in terms of colour and flavour. The deep crimson colour of its flesh is unique amongst hardy root vegetables; most are pale in colour, and the slightly sweet flavour, allied to its juiciness, makes it difficult to associate with the majority of savoury vegetable dishes. In the 16th and 17th centuries, this would have been no problem; much food was then a mixture, what one might call now 'sweet-sour', and the separation into savoury and sweet or dessert courses did not come until some hundreds of years later.

Varieties

Garden beetroot is closely related to sugar beet, which is simply a strain of *B. vulgaris* 'Crassa', developed and maintained over the years to provide the maximum amount of sugar when refined. It is closely related to Swiss Chard (seakale beet), grown for its stalks and spinach-like leaves. Besides the crimson form, there is a golden-rooted kind and another with white flesh. Globe-shaped beetroot are usually preferred, though there are cylindrical and long kinds.

Culinary uses

Beetroot is ill-treated in Britain and many other countries where it is traditionally served doused in vinegar and as a strident salad accompaniment to cold meats. Northern European countries have handled its sweet and subtle flavour in more interesting ways. Bortsch in its various forms is one example or serve it hot after boiling or baking in foil, the slices toasted in sour cream and chives. The skin should be unbroken so that the beetroots do not bleed and lose their colour; to test for tenderness, rub off the skins, do not prick with a fork. If the skin comes off easily, cooking is finished. Boiling can take from ½–2 hours.

Cultivation

Although it is a native of southern Europe, beetroot can easily be grown in cool temperate climates. The root, produced in the first of its two years, is the part required, so can be grown during the summer and dug up for storage in autumn. However, with protection such as cloches, straw or bracken, it will see the winter through in warm gardens.

It has a tendency to bolt, so choose bolt-resistant varieties wherever possible and water well in hot, dry weather. Each 'seed' is in fact a capsule of several seeds, and as each is comparatively large, it is easily station sown, and thinned at each station.

CULTIVATION

Site and soil sun; moist soils with reasonable drainage, manured for previous crop; not acid, prefers sandy fertile one

Soil preparation dig 1–2 spades deep, depending on variety to be grown; add lime in winter if needed; fork in compound potash-high fertilizer in spring if soil very light

Sow outdoors mid-spring, for early summer crop, must be bolt-resistant, station-sow 25mm (1in) apart, rows 20cm (8in) apart; sow late spring–early summer for maincrops and storage, sow as above but 7cm (3in) apart. Sow all 2.5cm (1in) deep

Thin thin early crops to 7cm (3in) apart; thin maincrops to 15cm (6in) apart

Care keep weeds under control; protect from birds as soon as seeds sown; mulch when plants large enough; water well in dry weather otherwise roots become woody, tasteless and cracked

Harvest pull up early beetroot when the size of a golf-ball; pull maincrops at cricketball size throughout summer, and pull remainder for storing from early autumn. Twist leaves off, do not cut from crown

Troubles mainly trouble-free; birds when young; mangold fly (leafminer) causes brown blistering in leaves

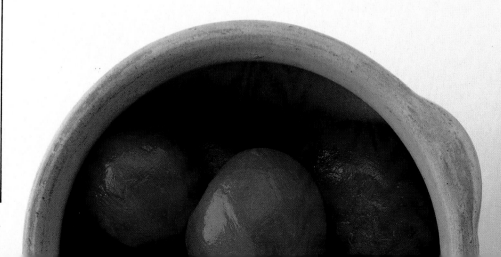

Type of plant biennial grown as half-hardy annual
Part eaten root
In season fresh, late spring–early autumn; stored, mid-autumn–early spring
Yield globe: 125g (¼lb) each long: 200g (7oz) each
Time from sowing to harvest round: 8–12 weeks, long: 15 weeks
Size round: 8cm (3¼in) diameter long: 15–23cm (6–9in)
Hardiness half-hardy
Seed viable 4 years
Germination period 15–24 days

RECIPE
Economical

Beetroot and apples
This makes a simple but delicious accompaniment to sausages, game, duck or ham.
500g/1lb cooked beetroot, sliced
1 large onion, sliced
50g/2oz butter
2 cooking apples, peeled, cored and sliced
salt and pepper
nutmeg
chopped chives or parsley
Melt the butter in a pan and cook the apple slices until tender but still firm. Remove from the pan and keep warm. Cook the onion until soft, add the beetroot and allow to warm through. Add salt, pepper and a little nutmeg and arrange in a serving dish with the apples. Sprinkle with chopped chives or parsley and serve immediately.

Brussels Sprouts

(Brassica oleracea bullata 'Gemmifera'*)* CRUCIFERAE

Brussels sprouts are very much a northern European vegetable, even a British one, though in Britain they were not commonly eaten until the middle of the last century. Similarly, they were mentioned in an American publication in the early 1800s, but were seldom grown in the US. Brussels, in Belgium, has to be their place of origin, though their birth is shrouded in mystery. No one seems to know for certain how or when Brussels sprouts first appeared, or even why a perfectly ordinary cabbage, or perhaps kale, produced buds turning into more small cabbages in the leaf-axils remaining after the main head had been removed.

Varieties

Brussels sprouts are round, compact buds of tightly packed leaves produced in the leaf-joints on the tall stem of a type of common cabbage. The earliest varieties produced widely-spaced sprouts; modern kinds develop them so closely packed as to be touching, and clothing the stem all the way up. There are early, midseason and late varieties, average height being about 75cm (2½ft). They are mostly green, but a few have deep red sprouts and red leaves.

Nutritional value

Brussels are amongst the most nutritious of the leafy vegetables: they contain a good deal of a variety of minerals and protein, a lot of carotene and vitamin C, and useful quantities of dietary fibre. Eaten raw, they are even more valuable.

Culinary uses

Normally eaten as a boiled or steamed side-vegetable, it is important that they should be only just tender, otherwise they become unpleasantly flavoured and a yellowish green colour, as well as soft and mushy. In France they are braised in wine. Small sprouts are much the best flavoured and this is where home growing scores so much, as the shop-bought kind are often large and beginning to open out. The tops of the plants can also be eaten, that is, the small cluster of loose leaves resembling a cabbage sold as 'sprout tops' after completion of harvesting a plant.

Cultivation

As with all the leafy brassicas, Brussels sprouts do best in well-firmed soil; if planted in loose, recently-dug soil the sprouts will have a tendency to 'blow', or open out to resemble miniature cabbages. They will be just as tiresome if supplied with too much nitrogen, so are preferably grown to follow another crop for which the ground was manured. They are not difficult to grow, though the effect of wind is a point to watch – supporting will be necessary for most varieties – and they need a long growing season. If planted after the end of midsummer, they will not make large enough plants before the winter to produce mature sprouts.

CULTIVATION

Site and soil sunny/open; sheltered from wind; deep, fertile, moist, alkaline, manured for previous crop
Soil preparation dig 2 spades deep, lime in winter, if necessary; rake in fertilizer 2 weeks before planting; firm soil by treading
Sow outdoors 1.5cm (½in) deep in nursery-bed early spring for autumn crop, mid-spring for early winter–early spring crop
Thin to 7cm (3in) apart when large enough to handle
Plant water site day before, if dry weather; dig up each plant when 3–4 leaves present, and straight stems, with large ball of soil, plant firmly in firm ground up to lowest leaf and water in; space 45cm (1½ft) apart each way dwarf vars., 75cm (2½ft) for remainder
Care keep well supplied with water after planting, if necessary daily in dry weather; watch for slugs/snails and birds; liquid feed in midsummer occasionally
Train/support if stems rocking in wind, mound soil round them a few weeks after planting; in early autumn tie to stakes to support against wind
Harvest from early autumn–early spring; cut or snap sprouts off, starting at bottom, remove blown sprouts, yellowing and lowest leaves; cut off top just before sprouts mature, but final sprout yield will be lower, or remove when all sprouts picked; 1 plant will crop for at least 8 weeks
Troubles slugs/snails; birds, caterpillars, flea-beetle, general brassica troubles; 'blown', sprouts, soft soil, too much nitrogen, too wet, not enough sun, loose planting

Citadel

Noisette

Bedford

ESSENTIAL FACTS 59

Type of plant biennial
Part eaten axillary bud, top leaves
In season early autumn–early spring
Yield ¾–1kg (1½–2lb)/plant
Time from sowing to harvest 28–36 weeks
Hardiness hardy
Seed viable 4 years
Germination period 7–12 days

RECIPE
Economical

Sprouts with chestnuts
Besides serving as a side vegetable, cooked as suggested earlier, small sprouts are delicious mixed with chestnuts peeled and boiled, tossed in melted butter, and served with turkey at Christmas time.

They can also be served with a cheese sauce, or braised with onions in a little stock.

Raw sprouts can be used in salads, provided the small ones are chosen, and then shredded; red sprouts would add colour to a salad made with Chinese cabbage and celery, and tossed in a vinaigrette containing hard-boiled eggs.

Cabbage

(Brassica oleracea) CRUCIFERAE

The original wild cabbage is native to the north Mediterranean and Adriatic coasts, and to southern England and Wales, but the modern cabbage is a hybrid descended from *Brassica oleracea* after centuries of breeding and selection. It is a member of the *Cruciferae* family and, if allowed to, would have yellow flowers in mid or late summer. Originally the plant produced a rather loose rosette of leaves at soil level, from which the flowering stems elongated the following summer, but by hybridising, the familiar tightly-packed heads of leaves, almost the size of a football, have been produced, and much the same shape, too.

Cabbages were known and grown by the Romans; they called them *caulis*, literally the stalk of a plant, but usually applied to cabbage-stems, and their plants were loose-leafed; cabbage derives from the French *caboche*, meaning head. *Caulis* eventually became 'cole' or the Scottish 'kale', and coleslaw or kohlrabi are both derivatives. Cato used cabbages, not just as food, but as a remedy for many illnesses and as a stimulant. The Greeks ate them, and they were certainly used as a food by man in prehistoric times throughout Europe.

Varieties

There are varieties to grow which mature in summer – all are round and green; others for winter use which may be dark green with smooth or blistered leaves – the latter are the Savoys – the red cabbage, available up to midwinter, and the kind with white hearts and mostly white outer leaves (often called Dutch cabbage in England). Then there are the spring cabbages, smaller with a conical centre, very welcome at that time of the year, both for a fresh green vegetable and for their sweeter flavour.

An oriental 'cabbage' which is increasingly being grown in Western gardens now is the Chinese cabbage, *Brassica siinensis*, very fast growing and ready in autumn and early winter. In appearance it looks more like a large cos lettuce, and has a crisp, light green leaf with a broad white midrib. Nevertheless one can weigh 1kg (2lb) when mature. With its ability to mature quickly, it is an excellent catch-crop, and also makes a good winter salad plant, if grown with the protection of cloche or greenhouse.

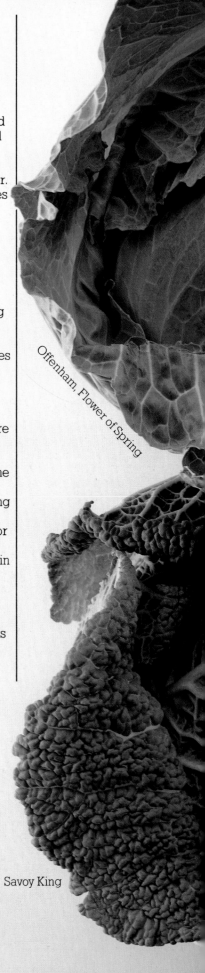

Offenham, Flower of Spring

Savoy King

CULTIVATION

Site and soil sunny, open; fertile, moist, neutral-alkaline, manured for previous crop

Soil preparation dig 1 spade deep in autumn; fork in lime in winter if soil acid; fork in general compound fertilizer 2 weeks before planting; firm the soil by treading before planting

Sow for cabbages cropping in summer, autumn and winter, and red cabbage, sow mid–late spring, but sow correct varieties for season; for spring-cropping, sow mid–late summer; sow thinly 1.5cm (½in) deep in seed-bed, space rows 15cm (6in) apart; for Chinese cabbage, sow mid–late summer where it is to grow at 10cm (4in) spacing, in rows 30cm (1ft) apart

Thin space seedlings 7cm (3in) apart; thin Chinese cabbage to 30cm (1ft)

Transplant plant in permanent positions when 3–4 leaves present; water plants the day before; plant firmly in weed-free but uncultivated soil, with lowest leaves just above soil surface; use straight-stemmed plants and water in after planting; plant summer, red and autumn cabbages late spring–early summer, winter cabbages midsummer, spring cabbage early–mid autumn; space at 30cm (1ft) each way for small compact heads, 45cm (1½ft) for large varieties; for spring cabbage allow 10cm (4in) between plants, 30cm (1ft) between rows

Care protect from birds, slugs/snails when young plants; water well in dry weather, especially Chinese cabbage, otherwise they bolt; liquid-feed once in mid-season if growth is slow; earth up stems of spring cabbages in late autumn; tie leaves of Chinese cabbage together with soft string

Harvest cut the cabbages off close to ground level when well-hearted; for spring and summer varieties and autumn cropping Chinese kinds, cut crosswise into the stump 1.5cm (½in) deep – second small cabbages will grow from stump; for remaining varieties dig up stump and roots and destroy; cut alternative specimens from spring cabbages before hearting, remainder will then grow larger

Storage red, and white winter varieties can be cut when ready before first bad frosts, the outer leaves removed, and stored in a dry, cool airy place until early spring. Chinese cabbage will store for a few weeks after cutting in a cool place, i.e. the salad compartment of a refrigerator

Troubles the most likely ones are: birds on seedlings, young plants, spring cabbage; snails/slugs; cabbage white butterfly; caterpillars; a variety of aphids; whitefly; club-root; cabbage root-fly; flea-beetle; soil-living caterpillars

ESSENTIAL FACTS

Type of plant biennial or short-lived perennial
Part eaten leaf
In season all year with right varieties
Yield red and green: ½–1½kg (1–3lb)/head spring: ½–¾kg (1–1½lb)/head white, Chinese and Savoys: 1–2kg (2–4lb)/head
Size all but Chinese: 30–45cm (1–1½ft) wide × 15–23cm (6–9in) high; Chinese: 23 × 30cm (9in × 1ft)
Hardiness hardy, except Chinese cabbage
Seed viable 4 years
Germination period 7–12 days

Holland late winter

RECIPE
Economical

Garbure
1 large green cabbage, sliced (6–8 leaves reserved)
500g/1lb carrots, sliced
500g/1lb potatoes, in chunks
2 leeks, sliced
2 large onions, sliced
2 shallots, chopped
3 small turnips, peeled and halved
100–150g/3–5oz dried haricot beans, soaked
2 cloves garlic, chopped
1 clove
bouquet garni
salt and freshly ground pepper
For the farci (stuffing)
6–8 large cabbage leaves
150g/5oz fresh breadcrumbs, soaked in milk
150g/5oz minced pork, chicken, duck or bacon
2 shallots, finely chopped
2 cloves garlic, finely chopped
1 onion, finely chopped
1–2tsp chopped parsley
salt and freshly ground pepper
Put all the soup ingredients into a large saucepan, cover with water and simmer gently for about 1 hour. Meanwhile, make the stuffing by combining all ingredients (except cabbage leaves) until well mixed. Blanch the cabbage leaves in boiling water for a few minutes until they become softened. Roll up knobs of the stuffing in the leaves and tie them with thread or fine string. Add *farci* to soup after 1 hour, and simmer for another 30–45 minutes before serving.

Nutritional value

Cabbages have good food values; they contain quite a lot of carotene and vitamin C, many mineral salts, dietary fibre, protein – a surprising amount of this in raw Savoys and other winter varieties – but have a low calorie count. Nutritionally, they are much better eaten raw, and of course the winter white cabbages form the basis for coleslaw, but red and Chinese cabbage make excellent raw salads. Provided the other types are shredded finely, selecting the inner leaves, they also can be used, mixed with such raw vegetables as sweet peppers, onions or celery.

Culinary uses

If they are to be cooked, steaming is preferrable to boiling. If boiling is unavoidable, use as little water as is possible without burning, cut the leaves finely, and cook until they are just tender. Otherwise, stir-fried, or used as a main course, stuffing whole cabbages or using individual leaves as the Greeks use their vine-leaves, are some of the alternative ways of serving them. Red cabbage is in a class of its own, and can be treated in all sorts of delicious ways.

Cultivation

The cultivation of cabbages is not difficult, but they are subject to a good many pests and diseases (see pages 42–43), and it is the combatting of these that can take much time and trouble. It was for the brassica family as a whole that rotation was invented, to avoid a build-up in the soil of such troubles as club-root, cabbage rootfly, or soil caterpillars; where club-root infects, ground should not be replanted with brassicas for at least eight years, as there is no cure.

But many of these problems appear because of wrong conditions and can be avoided if the crop is supplied with deep, well-cultivated neutral-alkaline soil, correct planting, choosing the right varieties for the time of year, and ensuring that there is always a good water supply. So many plants are attacked by insect pests in summer if they are short of water, and close spacing makes matters worse. Firm, solid-hearted cabbages will be obtained from firm soil, not recently dug or recently dressed with organic matter, and size of head is controlled to a large extent by the spacing.

Chinese cabbages are mostly troubled by slugs and snails; the general run of cabbage problems are not nearly so prevalent on them, and the water requirement, although great, is usually supplied by the weather for most of their life.

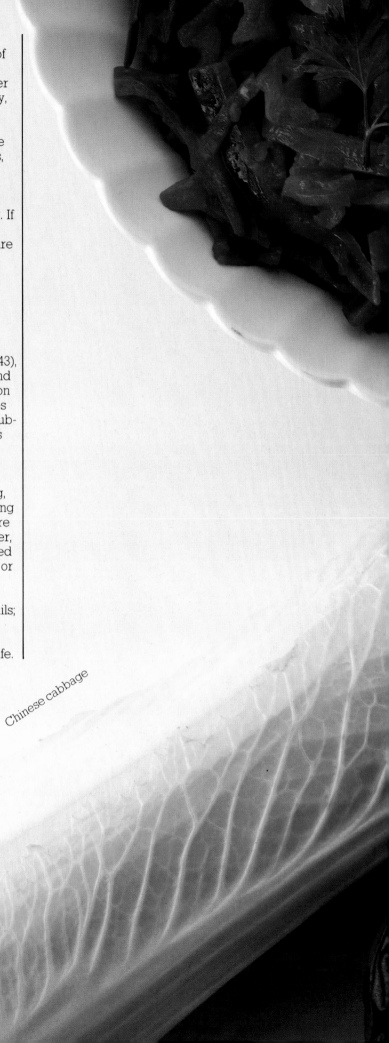

Chinese cabbage

RECIPES
Healthy

Red cabbage

500g/1lb red cabbage, shredded
1 cooking apple, peeled and
*   chopped*
2tbls brown sugar
1 medium onion, sliced
salt
black pepper
1tbls vinegar
2tbls oil
1tbls water
Cook the onion gently in 1tbls oil
until tender, add apple and soften,
then add sugar and stir the mixture.
Add cabbage and remainder of oil,
stir to ensure the cabbage is coated
in oil, then add vinegar, water and
seasoning. Turn all into an
ovenproof dish and cook with the lid
on in a slow oven, 150°C (300°F), gas
mark 2, for about an hour, or simmer
slowly on top of the stove in a
saucepan with a tight-fitting lid.

Exotic

Stir-fried Chinese cabbage

500g/1lb Chinese cabbage, cut into
*   5cm/2in strips*
30ml/2tbls corn oil
1 clove garlic, finely chopped
2 thin slices ginger, finely chopped
½tsp salt
toasted sesame seeds, to garnish
Heat the oil in a wok to a medium
temperature. Add garlic and ginger
and stir-fry quickly for a minute or
two (do not let them brown); add
Chinese cabbage and salt and cook
for 3–5 minutes, stirring constantly
until cabbage is cooked but still
crunchy. Serve immediately
garnished with lightly toasted
sesame seeds and a sprinkling of
sesame oil, if desired.

Red cabbage

Calabrese

(American green sprouting broccoli, *Brassica oleracea* 'Italica') CRUCIFERAE

The cabbage family is large and varied, and a comparatively new addition to it has come from Italy in the shape of calabrese, named after the district, Calabria, where it has been grown since mediaeval times. Calabrese is essentially a type of green-flowering sprouting broccoli, but could be regarded as a halfway house between that vegetable and cauliflower, as it consists of a tightly packed head of flowers at the top of the stem, not unlike the curds of cauliflower followed, after cutting, by the production of edible flowering shoots similar to the florets of sprouting broccoli, in the leaf joints lower down the stem.

However, unlike purple-sprouting broccoli, it is tender, and crops in summer and autumn; a further difference is that it can be sown where it is to grow. Finally, it grows into a much shorter plant, and the curd or florets are a beautiful bluish-green colour. In markets and shops it is frequently sold as broccoli.

Nutritional value

Nutritionally, calabrese has slightly less dietary fibre, protein, water and calories than cabbage, and more in the way of vitamins and mineral salts than cauliflower. Although it starts to mature in late summer, when many other more traditional vegetables are available, it will continue to produce florets until the really cold weather, i.e. until mid autumn at least, and often into late autumn. It is not unknown for florets to be picked right through winter in sheltered gardens or mild winters, so it can be well worthwhile growing; any summer surplus can be frozen or pickled.

Culinary uses

The more delicate flavour of calabrese, together with its crisp texture, makes it a more versatile vegetable for culinary use than sprouting broccoli; besides its place as a side vegetable, steamed or boiled, it has enough flavour to merit its use as a starter served, like asparagus, with simply melted butter or hollandaise sauce and for inclusion in salads cooked or raw and served hot or cold. The crisp florets are also widely used in Oriental stir-fried dishes.

Cultivation

Cultivation follows the same pattern as for the general run of brassicas, with the emphasis on protecting it from cold in the early stages, and keeping it growing fast. It does better if planted where it is to grow, when it crops more quickly and more heavily. It does not take up as much space as sprouting broccoli, nor does it occupy the ground for anything like as long. See pages 42–43 for pests and diseases that can affect brassicas.

CULTIVATION

Site and soil sun/open; most soils, preferably moist, not acid, manured for previous crop

Soil preparation dig 1 spade deep in winter, add lime if acid; soil should be firm for planting

Sow early–mid spring in seedbed, with cloche protection; late spring where it is to grow; protect at night; sow 1.5cm (½in) deep

Thin to 7cm (3in) apart); for permanent plants to 30cm (1ft) final spacing

Plant when 3–4 leaves present, protect until end of late spring, plants 30cm (1ft) apart each way; move with good ball of soil, water the night before, and water in after planting, plant firmly up to lowest leaves

Care keep free of weeds, slugs/snails and birds; water well early in dry weather, mulch; liquid-feed occasionally if soil poor

Harvest cut terminal head when full-grown and tightly packed, before opening out; snap off florets which form later when about 7cm (3in) long and florets still in bud; do not strip completely and do not allow florets to flower, otherwise production stops

Troubles general cabbage troubles; watch mainly for slugs/snails, birds, caterpillars, flea-beetle

ESSENTIAL FACTS

Type of plant annual
Part eaten flower-head, florets
In season late summer–mid autumn, sometimes longer
Yield about ½kg (1lb)/plant
Time from sowing to harvest 12–16 weeks
Size plants 60–75cm (2–2½ft)
Hardiness tender
Seed viable 4 years
Germination period 7–12 days

RECIPE
Exotic

Indian braised calabrese
1kg/2lb calabrese
1tbls oil
2 teaspoons yellow mustard seed
1in piece of fresh ginger, peeled and finely chopped
1 clove garlic, finely chopped
½ teaspoon turmeric
2 teaspoons cumin
pinch of cayenne pepper
salt and pepper
1tbls shredded green pepper
coriander leaves, to garnish

Cut the calabrese into small florets. Heat the oil in a saucepan, add the mustard seeds and when they start to pop, add the ginger and garlic. Cook gently until soft but not brown. Add the turmeric and stir over low heat for about 1 minute. Add the calabrese florets, stir well and then add cumin, cayenne pepper and salt and pepper to taste. Add 2tbls water, cover the pot and cook gently for about 5 minutes. Then add the green pepper, stir, cover and cook for a further 5 or 6 minutes or until the calabrese is tender and the liquid absorbed. Garnish with coriander leaves and serve.

Carrot

(Daucus carota) UMBELLIFERAE

Carrots are root vegetables, with a swollen bright orange root, and an attractive cluster of delicate, ferny leaves up to 20cm (8in) long, both produced in the first year of growth.

White or beige-rooted carrots can still be found growing wild in Europe, and although the carrot was included in lists of medical plants by the Greeks and Romans, it does not seem to have been used as food in Europe until the 14th century. They were not introduced to Britain in any quantity until the Flemings arrived during the reign of Elizabeth I. Later, ladies of Charles I court used to wear the foliage as a decoration.

Varieties

Available in many varieties, the modern carrot's shape is long and tapering, finger-like or stump-rooted and they are classified in three categories: long, intermediate and short – or stump-rooted. The most useful for the gardener and cook are the intermediate types, as they can be grown to provide small succulent roots early, and also maincrop roots later for storing. The stump-rooted are early varieties with a delicious flavour, to be eaten as soon as they are ready; the long kinds are often grown for showing as they can reach enormous lengths, 45cm (18in).

Nutritional value

The highest mineral content is that for potassium; there are large amounts of carotene (vitamin A) – 500g/1lb raw carrots contains 4½ times the body's daily requirement – and a good deal of vitamin C. Dietary fibre is fairly high, and there are 23 calories/100g, but there is little protein, and only a trace of fat.

Culinary uses

The texture of raw carrot is crunchy, the flavour slightly sweet, and they were formerly much used for sweet dishes, such as carrot jam or carrot tarts. Nowadays, they are generally regarded as a vegetable, usually boiled or steamed, but if grated they are a valuable salad ingredient, and simmered for Crécy soup, they are delicious; sliced vertically they also make excellent crudités for dips.

Cultivation

Provided a sunny situation and well-drained fertile soil free from stones is available, carrots are not difficult to grow, but do have one pest which is difficult to deal with, and can cause considerable damage. This is carrot fly, whose maggots burrow in to young roots and feed, destroying the quality and providing a foothold for other pests, and for fungus disease. The adult flies appear from late in mid-spring, with a second generation in late summer–early autumn, and lay eggs a few days later in the soil close to the crown of the carrots. Infested roots develop reddish tints in the leaves, wilting in hot weather, and later turn yellow, and collapse completely. Either treat the soil with an insecticide before sowing, sow the seed amongst garlic or onion, or delay sowing until late spring to midsummer. Sow thinly, thin the young plants in the evening, and dispose of the thinnings at once deep in the compost heap as the fly is attracted by the smell of the bruised foliage. Protect with cloches at time of egg-laying.

CULTIVATION

Site and soil open/sunny, medium-light soil, ideally sandy/fertile, no stones, manured for previous crop
Soil preparation dig 1–2 spades deep, depending on carrot type, fork in compound potash-high fertiliser 10 days before sowing if soil known to be short of nutrient-prepared surface for sowing
Sow stump-rooted, outdoors early spring–midsummer, thinly 1.5cm (½in) deep in rows 15cm (6in) apart; intermediate and long, outdoors mid-spring–early summer, thinly, 1.5cm (½in) deep in rows 20–30cm (8–12in) apart
Thin stump-rooted to 5cm (2in) apart, in 2 stages; intermediate and long, to 10–15cm (4–6in) apart, in 2 stages
Care weed carefully while carrots young, do not bruise foliage, firm soil round carrots immediately after thinning and water lightly; mulch on light soil; watch for greenfly in drought; carrot-fly from late spring
Harvest use thinnings in salads if large enough late spring–early summer; dig up stump-rooted and use at once, late spring–late summer; dig up remainder and use fresh late summer–mid-autumn, *OR* leave through winter, protect, *OR* dig up and store in dry sand or peat boxes in shed for winter
Troubles greenfly; carrot fly; cracking due to drought; forked roots due to large stones/organic matter

Type of plant biennial grown as annual
Part eaten root
In season fresh, late spring–mid-autumn; stored late autumn–early spring
Yield 3–4kg (6½–8½lb)/1m (3ft) row, slightly less for early crops
Time from sowing to harvest early/salad carrots, approx. 10 weeks; maincrops, 14–18 weeks
Size stump or short-rooted (round or finger-like), 5–12cm (2–5in) long; intermediate, 20–23cm (8–9in); long, 30cm (12in) or more
Hardiness hardy
Seed viable 4 years
Germination period 14–24 days

RECIPE
Exotic

Glazed carrots
500g/1lb carrots, trimmed and scraped
50g/2oz butter
1tbls sugar
fresh chopped parsley, to garnish
Put the carrots whole in a pan of lightly salted boiling water, cover and simmer for about 15 minutes or until just tender. Drain. Add the butter and sugar to the pan, then the drained carrots. Cook over gentle heat, turning the carrots, until evenly glazed. Sprinkle with chopped parsley, to serve.

Cauliflower

(Brassica oleracea 'Botrytis') CRUCIFERAE

The closely-set white curds of cauliflower have been a familiar part of European diet for about 500 years. The Romans knew and relished them, calling them 'cauliflora', i.e. a flowering cabbage, and cauliflowers were cultivated in Spain in the 12th century at about the time that hybridizing, in Cyprus and Italy, was producing better and bigger cauliflowers. The vegetable seems to have made its way to France from the eastern Mediterranean, and from thence as an import to London in the early 16th century.

Varieties

From this it will be seen that the cauliflower is not as hardy as all that, hence its reputation for being difficult to grow. Winter-maturing varieties are, technically, broccoli, though usually regarded simply as winter cauliflower. For summer cropping, there is a separate collection of varieties and yet another which head up in autumn, and this group includes the Australian kinds – amongst these there is one with purple florets, which cook green.

Nutritional value

Amongst the vegetables, its food value is comparatively low, vitamin C being the only constituent of any value. The calorie value is low, a point in its favour, but so also is the dietary fibre and mineral content. Surprisingly, it contains more protein than potatoes.

Culinary uses

There are a good many culinary uses for cauliflower, apart from the traditional boiling or steaming; the latter method results in a better flavour. Individual florets can be added raw or cooked to salads, used as crudités, or as an hors d'oeuvre with dressings or mayonnaise. It can be served with sauces, it can be made into a soufflé, pickled in a number of ways and used in soup.

Cultivation

Like lettuce, the right variety has to be grown for the right season, to ensure good tight heads of firm white curd. A further requirement for successful cauliflowers is planting in firm ground – it should not have been recently dug or manured, otherwise the heads will be loose and open. Finally a sandy or gravelly soil is quite unsuitable for cauliflowers, especially in areas of low rainfall. Growing a cauliflower crop to a successful conclusion is a great test of gardening skill, but it has to be said that growing a row of perfectly formed round heads has its disadvantages, since thay all form at once. To avoid this, one summer and one winter variety sown two or three times each, in succession, will supply continuity without glut.

CULTIVATION

Site and soil sunny, not a frost-pocket; deep, rich moisture-retentive soil, not acid or strongly alkaline

Soil preparation dig 2 spades deep in autumn; mix in rotted organic matter for summer varieties (winter varieties best on ground manured for previous crop); lime if necessary in winter; apply fertilizer 2 weeks before planting, raked in

Sow summer vars., early–mid-spring; autumn vars., mid–late spring, and winter–spring vars., late spring. Sow outdoors 1.5cm (½in) deep in a seed-bed, spaced singly about 25mm (1in) apart, in rows 15cm (6in) apart

Thin to 7cm (3in) apart in plenty of time to prevent them becoming weak and straggly

Plant move to their permanent places when 3–4 leaves present; water the plants the day before, dig up with as much soil round the roots as possible, and plant in firm soil, with the lowest leaves just above the soil surface. Water in well. Space summer and autumn varieties 60cm (2ft) apart each way, winter varieties 75cm (2½ft) apart; 15cm (6in) for small summer vars.

Care keep free of weeds; protect against birds and slugs/snails when young; always water well in dry weather; ridge up winter cauliflowers in autumn round the stems to prevent wind-rocking and help excess winter rain drain off; break inside leaves over summer vars. to protect from hot sun; same for winter vars. to protect from frost

Harvest cut heads when usable before full size to avoid glut

Store dig up with roots on and hang summer vars. head downwards in a dark cool place, spraying occasionally, when they will keep a week; winter vars., if planted in boxes, will keep a month

Troubles general brassica troubles; birds; slugs/snails; caterpillars; flea-beetle; small or non-existent heads: dry soil, check at planting due to cold, drought, transplanting too big or too small, too long in pots or blocks, shade, 'blind' plants (no growing tip at planting time); lack of trace element molybdenum produces narrow leaves, no heads (whiptail), due to acid soil; brown curd, distorted leaves, boron deficiency, due to high alkalinity

Type of plant biennial
Part eaten florets
In season late winter–late autumn
Yield ½–1¼kg (1–2lb) per plant
Time from sowing to harvesting
16 weeks summer vars., 24 weeks
autumn vars., 40 weeks winter vars.
Size whole plant: 17–25 × 15–38cm
(7–10 × 6–15in)
Hardiness slightly tender
Seed viable 4 years
Germination period 7–12 days

RECIPE
Exotic

Curried cauliflower
1 medium-sized cauliflower
2 sticks celery
250g/½lb tomatoes
125g/¼lb French beans
275ml/½pt/2 cups stock
2 medium-sized onions
1 garlic clove
salt and black pepper
2 tbls mild curry powder (or
according to taste)
oil
Divide the cauliflower head into
bite-sized florets; chop the celery,
slice the tomatoes, French beans
and onions; place all except the
onions in a dish, and then sprinkle a
mixture of the salt, black pepper
and curry powder evenly over
them. Sauté the sliced onions in the
oil until just tender, add the curried
vegetable mixture to them together
with the garlic, crushed, and pour
the stock over. Simmer gently
uncovered for about 20–30 minutes,
or until tender, when most of the
stock will have been absorbed.

Celeriac

(Knob celery, turnip-rooted celery, German celery, *Apium graveolens*) UMBELLIFERAE

Celeriac is closely related to celery; both are varieties of the same species, but celeriac is grown and eaten for the swollen rootstock just below the crown, forming a large rounded globe, from which the true roots extend. It has a cluster of green leaves on top of green, slightly fleshy leaf stalks. The flavour of the 'root' is exactly like celery, but the leaf-stems are never eaten as they taste bitter and are tough and stringy. Basically its history is the same as that of celery and, though it has never been grown in such large quantities, it has always been popular in Europe as a cooked or salad vegetable.

Culinary uses

To prepare it for eating the fibrous roots on the outside should be removed and the skin peeled off. It can then be cut into chunks and boiled or steamed until tender, or grated for salads, sprinkling it with a little lemon juice to prevent browning. Other uses include soup, a starter, and inclusion in casseroles. In France, it is widely served as an hors d'oeuvre, *céleri-rémoulade*, flavoured with mustard mayonnaise.

Nutritional value

Nutritionally it has one of the highest potassium contents of the vegetables, quite a lot of dietary fibre, a useful amount of vitamin C, and some protein, together with other mineral salts and a moderate amount of calories.

Cultivation

It is a versatile vegetable which supplies an authentic celery flavour without the trouble of growing celery itself, a time-consuming plant which needs careful soil preparation and propagation. Celeriac is easily grown, provided it has plenty of water early in its life, and needs little attention while growing.

CULTIVATION

Site and soil sunny/open; most soils, preferably moist and fertile
Soil preparation dig 1 spade deep in early winter and mix in organic matter
Sow mid spring indoors; sow 0.6cm (¼in) deep in 5cm (2in) peat pots or soil blocks; maintain 16°C (60°F)
Thin to 1 seedling in each pot
Plant harden off and plant outdoors late spring–early summer; space plants 30 × 45cm (1 × 1½ft); plant pot/block complete, with swollen stem-base at soil level; water in
Care keep well-watered in early stages, at any time whenever drought occurs; watch for slugs on young plants; liquid-feed weekly from midsummer on average–light soil until early autumn; remove sideshoots; cover crown with soil in early autumn
Harvest start to dig as required in late autumn, and throughout winter
Store dig up in late autumn and store in damp sand for winter if garden a cold one
Troubles slugs/snails; celery fly; mice/voles eat in ground in hard winters; small 'roots', lack of water

ESSENTIAL FACTS

Type of plant biennial
Part eaten swollen stem-base, 'root'
In season late autumn–late winter
Yield average 250g (½lb)/plant
Time from sowing to harvest 24–26 weeks
Size plants 38 × 38cm (15 × 15in) 'root', about 10cm (4in) wide
Hardiness hardy; seeds need little warmth
Seed viable 6 years
Germination period 10–20 days

RECIPE
Economical

Celeriac soup
250g/½lb celeriac, peeled and diced
1 medium onion, sliced
300ml/½pt stock
300ml/½pt milk
25g/1oz butter
lemon juice
Sprinkle the celeriac with lemon juice immediately it is peeled and chopped to prevent browning. Melt the butter in a pan, and add the vegetables, cook gently until transparent. Then add the stock and milk and simmer until the vegetables are soft. Liquidize, reheat and serve with croûtons and snipped chives sprinkled on to the surface.

Celery

(Apium graveolens 'Dulce') UMBELLIFERAE

Celery is very much an Italian vegetable, and it was in Italy that its culinary use really started, during the Middle Ages, and it was where most of the breeding and improvement was carried out to produce the plant we have today. Celery was regarded as a medicinal plant by the Romans and the Greeks, and in classical times was also used in funeral wreaths and to crown the heads of athletes.

The wild species is a native of marshy and salty regions, chiefly in Europe, Africa and North and South America, though it was mentioned by the Chinese writers of the ancient dynasties, and has been found in the Himalayas. In Britain wild celery can still be found in moist ground, close to the sea. Successful cultivation therefore suggests plenty of water and generous additions of rotted organic matter, seaweed being a particularly good form.

Varieties

Today there are several different kinds of celery regularly grown, the two main groups being self-blanching celery, and winter celery. The former is naturally pale in colour and, provided it is grown closely, and black polythene sheet is provided to cut out light even more, will be blanched with little further trouble, to be cropped in summer and early autumn. Winter celery needs to be blanched by surrounding individual plants with collars, and then earthing up the row, when it will be ready for use in winter. It is taller and larger than summer celery. Within these two groups there are varieties with red or pink stems, and with green stems; all have the authentic celery flavour, and are crisp and succulent when properly grown.

CULTIVATION

Site and soil sunny, open; rich moist soil, high humus content

Soil preparation winter celery: dig a trench 38cm (15in) wide and 2 spades deep in mid-spring, fork up base of site, add layer of very well-rotted organic matter, trodden down, return soil until 7cm (3in) short of soil surface; self-blanching celery: dig 1 spade deep in mid-spring and mix in rotted organic matter generously

Sow under cover in mid-spring in 5cm (2in) peat pots or in soil blocks 0.6cm (¼in) deep, maintain 16°C (60°F)

Thin to one plant in each pot when large enough

Plant outdoors when 5–6 leaves present; harden off first; do not plant too deeply; space winter celery 23cm (9in) apart, and fill space at top of trench with water after planting; space self-blanching celery 23cm (9in) apart each way, planted in a square so plants are automatically blanched. Plant giant and green vars. 30cm (12in) apart; water in

Care watch for slugs/snails at all times; water regularly and generously at all times, especially in drought; liquid-feed weekly from the middle of early summer to harvesting or early autumn, whichever is earliest

Blanching surround outside block of self-blanching celery with black plastic sheet in midsummer; when winter celery is 30cm (1ft) tall remove sideshoots or leaves, tie stems together just below top leaves and wrap newspaper or corrugated cardboard round stems, ensure no slugs in celery centre, then pile soil round the plants to 15cm (6in) depth; repeat late in the summer, and a final time in early autumn so that only the leaves show, and the soil slopes steeply to ground-level; prevent soil falling into the plants' centre

Harvest dig self-blanching celery late in midsummer as required, from outside first; dig winter celery from mid–late autumn

Troubles slugs/snails especially on young plants and in hearts; celery fly on leaves, produces pale brown blisters; heart rot, injury due to frost, slugs or cultivation; bolting, lack of water, cold at planting, or seedlings too large at planting; stalks splitting, drought or too much nitrogen

ESSENTIAL FACTS

Type of plant biennial
Part eaten stem
In season midsummer–midwinter
Yield ¾–1kg (1½–2lb)/plant
Time from sowing to harvest summer var. 14 weeks; winter var. 26 weeks
Size 30–60cm (1–2ft) tall
Hardiness summer, tender; winter, hardy, but seed needs protection
Seed viable 6 years
Germination period 18–28 days

RECIPES
Healthy

Fennel and celery with blue cheese dressing: slice thinly equal quantities of celery sticks and Florence fennel, coat with a vinaigrette into which a few tablespoonfuls of blue cheese have been blended, and mix both vegetables together. Serve garnished with fresh dill.

Stuffed celery sticks: slice celery sticks into 15mm (½in) pieces, put in cold water for ½ hour, remove and dry. Mix with chopped hard-boiled egg (1 egg per stick) and vinaigrette made with garlic, serve sprinkled with chopped parsley and thyme.

Waldorf salad: use equal quantities of diced celery and apple (Cox's Orange Pippin or a russet), add pieces of walnut, and mix with mayonnaise. Serve chilled.

Nutritional value

Nutritionally, celery has a good deal of vitamin C in it, making it a particularly useful vegetable in winter; unusually there is a great deal of chlorine and sodium in it, one reason why it does well in coastal regions, since sea-salt is sodium chloride. There are few calories but moderate amounts of protein and dietary fibre.

Culinary uses

Its crisp texture when raw ensures its popularity as a salad vegetable, and for keeping teeth in good condition, and its flavour is acceptable in all kinds of recipes, particularly vegetarian ones, in which it helps to lift what might otherwise be rather blandly-tasting dishes. Eaten raw, it can be used as part of a crudités dish; it can be combined with cheeses, ham or asparagus as a starter, or for cocktail nibbles, and chopped up for use in many, many salad recipes. As a vegetable it can be boiled or steamed; braised, with butter and parsley, it is delicious; it makes an excellent soup on its own, and can be added to all kinds of casseroles, particularly chicken or pork, or blended with other ingredients to form delicious sauces.

Cultivation

Its cultivation is complicated, and takes more time than most vegetables. Winter celery is the more protracted of the two as it takes much longer to mature; it needs care in soil preparation, propagation and planting, and has to be blanched, but once ready in autumn, can be harvested until mid winter. Summer celery is ready in a few months from sowing and some varieties can be harvested until mid autumn. It needs much less care in digging and planting, but does do much better if grown in a frame; without this, some method of blanching the outside plants will be necessary.

CULTIVATION

Site and soil sun/open; most soils
Soil preparation dig 2 spades deep, use ground manured for previous crop
Sow green salad-leaved vars. midsummer; red-leaved: early–late summer, forcing vars: early summer; sow thinly outdoors, rows 30cm (1ft) apart, seed 1.5cm (½in) deep
Thin to 15cm (6in) apart forcing and red-leaved vars., remainder 20cm (8in)
Care watch for slugs/snails, keep free of weeds; water well in dry weather, especially at sowing time
Harvest cut heads from Sugar-loaf type in autumn, take single leaves from red vars. in summer but not too many, harvest heads as they mature in autumn
Protect cover red chicory with cloches in autumn/winter, in cold and/or wet weather
Blanch dig up Witloof chicory in late autumn when weather cold, trim off remains of leaves and tips of roots; choose roots about 4cm (1½in) wide at crown, reject forked roots, trim root back to 15cm (6in) long. Store horizontally in boxes of moist peat or sand. Force a few at a time by placing 4 upright in 20cm (8in) pot of moist compost, with crowns just exposed, and covering with another inverted pot, with blocked-up hole; keep at 13°C (50°F) in the dark; cut in 3–4 weeks
Troubles slugs/snails; soil-living caterpillars

Chicory

(Endive, Witloof *Cichorium intybus*) COMPOSITAE

Chicory is not grown nearly enough in the English-speaking countries; in Europe it is a regular part of the salad scene, especially in northern Italy where red chicories are widely grown. In Britain, it is mostly known only for the chicons, the forced blanched shoots sold in winter but the green lettuce-like leaves make perfectly good salads in autumn with the minimum of blanching, and the red winter chicories do not even need this, and can be used all through winter and into spring.

The wild species of chicory grows throughout Europe, the United States, north Africa and the western part of central Asia. Selected strains of this, which have large roots, are the chicories grown to provide coffee substitute in the form of the ground-up root. The leaves were blanched for salads by the ancient Egyptians; the plants were also used medicinally by the Romans, but chicory's main use for thousands of years has been as a vegetable.

Varieties

The green leaf chicories fall into two groups: the Sugar-loaf variety, which looks much like a large, rather open cos lettuce, and the Witloof chicory, grown for blanching in winter. The Sugar-loaf chicory is mostly to provide heads like lettuces in autumn, lasting into winter with protection, but can also be grown to provide small leaves from young plants before they mature. Witloof chicory has much longer narrower, slightly lobed leaves, too bitter to eat, and the root has to be dug up in autumn and forced to produce a second crop of leaves, tightly wrapped and white, which form the familiar chicons.

The red-leaf varieties such as *radicchio* are much more like lettuces but have the great advantage that they are hardy and will stand through the winter. Indeed, cold only improves their flavour, and it certainly intensifies the red speckling, tinting and veining typical of these varieties. As a basis for raw winter salads they have no equal, and their crispness and sharp taste leaves the average winter cabbage lettuce standing.

Nutritional value

Nutritionally chicory is similar to lettuce but better, though the chicons are not as valuable as the leaves. They contain mineral salts, and vitamin C, have a low calorie and carbohydrate value – some of the latter is inulin – and little dietary fibre.

Culinary uses

In the kitchen, the leafy salad kinds can be introduced to a variety of salads to lend them a distinctive flavour, which blends well with vinaigrettes, mayonnaise and other dressings. The chicons can be eaten raw, or braised and served with different sauces, and stuffed, curried or baked with ham.

Cultivation

Cultivation is no more complicated than for lettuce; sowing times are much later, in summer or early autumn, and blanching is necessary for some varieties, for which there are several techniques.

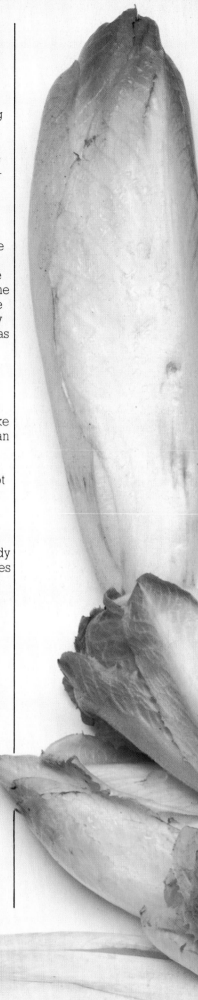

75

ESSENTIAL FACTS

Type of plant perennial
Part eaten leaf, root
In season early autumn–mid-spring
Yield 15–20 heads/3m (10ft); each 90–150g (3–5oz)
Time from sowing to harvest mid-autumn–mid-spring
Size 15–23cm (6–9in) x same; chicons 12–20cm (5–8in)
Hardiness hardy
Seed viable 6 years
Germination period 6–12 days

RECIPE
Economical

Chicory with bacon and cheese
4 chicons
2 slices smoked back bacon
1 medium onion
50g/2oz fresh brown breadcrumbs
2tbls parsley, chopped
50g/2oz grated Cheddar cheese
1 garlic clove
40g/1½oz butter
milk
salt and black pepper
Simmer the chicons for about 8 minutes, then remove and drain. Chop the bacon, onion and garlic, and mix this with the breadcrumbs soaked in a little milk, parsley and most of the cheese, reserving a little for garnishing. Season with salt and pepper to taste. Grease an ovenproof dish with a little butter, put half the bacon and cheese mixture in the bottom, then the chicons side by side, then the rest of the mixture spread evenly over them. Pour the rest of the butter, melted, over the stuffing and bake at 190°C/375°F/gas mark 5, for 30 minutes. Sprinkle the rest of the cheese over the top and brown under the grill.

Cucumber

(Cucumis sativus) CUCURBITACEAE

Cucumbers are tropical plants, thought to have originated in southern Asia, and trail or climb by means of tendrils twined round the nearest support.

The history of the plant is a long one, reaching back more than 3,000 years. It was grown extensively in the Orient, especially in India and tropical Asia. From there it spread to ancient Egypt and the Greeks and Romans; one of the Roman emperors, Tiberius, ordered that cucumbers should be available throughout the year and in the first century A.D. they were grown under cover of specularia, buildings with thin sheets of mica doing duty for our modern glass, to let in light and keep the plants warm. The Romans introduced them to Britain and they were part of a list of vegetables eaten cooked, described in the 12th century.

Varieties

There are two groups: one produces long smooth fruits, without being pollinated, and must be grown indoors; if it is pollinated, the fruits are bitter and become swollen at one end. The other develops comparatively short, thick fruit with a bumpy skin – these are ridge cucumbers which can be grown to maturity outdoors in cool temperate climates; the female flowers of these must be fertilized to produce cucumbers. The indoor cucumbers are prolific in their production of stems, shoots and leaves, and one plant can bear at least 25 fruits; the ridge varieties also grow strongly, with many sideshoots, and are grown trailing along the soil or up supports; they may set nearly as many fruits.

Nutritional value

Nutritionally, cucumber's main claim to fame is its vitamin C content; otherwise, although it contains a range of minerals, protein and fat, the amounts are low, as is the calorie value. As a thirst quencher, it is ideal – it is almost completely water. If eaten with the skin left on, there is some dietary fibre content.

Culinary uses

Nowadays, cucumbers are nearly always eaten raw, in salads, though cucumber soup is delicious, iced and served with sour cream. Chunks dipped in batter and deep-fried or cucumber braised in butter or simmered and added to mousse are more ways of cooking it. The tiny cucumbers called gherkins can be pickled or used in chutney.

Cultivation

Cultivation of ridge cucumbers is not difficult, given a good summer in cool temperate climates. Greenhouse cucumbers need careful attention to soil preparation, temperature and humidity maintenance, training, and removal of male flowers, unless an all-female variety is grown. Both kinds need copious quantities of water.

CULTIVATION

Site and soil ridge: sun, shelter; moist, fertile, well-drained; indoor: some shade, soil as for ridge

Soil preparation dig holes 30 × 45cm (1 × 1½ft), put mixture of rotted organic matter and potting compost in them, to fill, and cover with low mound of topsoil; 4–5 plants sufficient for a family; do this 2 weeks before planting; space holes 75cm (2½ft) apart for ridge, 45cm (1½ft) for indoor

Sow ridge: indoors late spring in 21°C (70°F); indoor: mid-spring; sow 2 seeds on edge or point downwards 1.5cm (½in) deep in 7cm (3in) pot

Pot ridge may need potting into larger pot before planting outdoors, disturb roots as little as possible

Plant ridge: outdoors early summer, protect from cold; indoor: late spring; plant root-ball without disturbance, water in

Care ridge: water copiously in dry weather; protect from cold until midsummer; watch for slugs/snails; liquid-feed from midsummer; mulch; place swelling fruits on tiles to keep clean; indoor: water heavily and maintain high humidity; keep ventilated; feed from time 1st fruits start to swell; watch for slugs/snails

Pollination do not allow indoor kinds to be pollinated, remove male flowers; ridge vars must be pollinated, insects will do it; female flowers have long swelling behind flower

Prune/train ridge: stop main stem when 7 leaves formed to encourage fruiting sideshoots, and stop these also at same place if no flowers; indoor: train main stem up wires to roof of greenhouse and stop; stop sideshoots at 2nd leaf beyond female flower; stop sideshoots without flowers when about 60cm (2ft) long; tie all stems to supporting grid of wires or canes

Harvest ridge: cut off stalks when 15–20cm (6–8in) long, before skin begins to yellow; indoor: cut when 30–38cm (12–15in) long; keep cutting both kinds to maintain production of more fruit

Troubles slugs/snails; mildew; collar rot; embryo fruit withering from tip, irregular water supply, cold, draughts, pruning too much, bad soil drainage; bitterness, pollination of greenhouse kinds, or heavy pruning, hot sun after cool period, temperature drop or sudden loss of water

Type of plant climber or trailer, annual
Part eaten fruit
In season indoor: midsummer–mid autumn; ridge: late summer–mid-autumn
Yield ridge: 15/plant; indoor: 25/plant
Time from sowing to harvest 12–14 weeks
Size fruit, ridge: 15cm (6in) long; indoor: 30–38cm (12–15in) long
Hardiness tender
Seed viable 7 years
Germination period ridge 6–14 days; indoor 3–5 days

RECIPE
Exotic

Dressed cucumbers
1 cucumber
125g/4oz cream cheese
125g/4oz soured cream
2 garlic cloves, crushed
125g/4oz smoked cod's roe
sprigs fresh dill
chives
coarsely ground black pepper, to garnish

Slice the cucumber into thinnish ovals. Place in a colander, sprinkle with salt and leave for about 45 minutes. In the meantime, prepare the two fillings. Firstly, in a food processor or blender mix the cream cheese, crushed garlic and half the soured cream together until well blended. Season to taste if desired. Set aside while you whisk the smoked cod's roe and remaining soured cream together until smooth and evenly blended. Season with a little lemon juice if desired. Wash the cucumber slices under running water to remove salt and dry each slice thoroughly with kitchen towel. Spread the cream cheese and garlic mixture on half the slices, and the cod's roe cream on the other half. Garnish with sprigs of dill and chopped chives. Add a little coarsely ground black pepper to each slice if liked.

Endive

(Batavia, chicory, escarole Cichorium endivia) COMPOSITAE

Endive was cultivated by the Romans and Greeks, probably for its medicinal use as well as for culinary purposes, when they ate it both cooked and raw. In France it was grown purely for medicinal use until well on in the Middle Ages; only after that did the French grow it as a food. In Britain it was being grown in the 15th century. However, it is still not as popular in Britain as on the Continent, where it is in daily use as a salad, particularly in winter; its mildly bitter flavour can be ameliorated by blanching.

Varieties

Endive is part of the same plant group as celery, and is a native of the same part of the world, in particular southern Asia and northern China. In France, one type is in fact called *'chicorée frisée'*. This is the kind with narrow, curled and much toothed leaves, looking like a curly green mop, and growing into a flattish rounded mound. The other sort is sometimes called Batavian endive, or escarole, from its much broader leaves, and is altogether a taller and larger plant, though still rounded. Neither produce the solid hearts that a cabbage lettuce does, but are a very useful autumn and winter salading. They are less prone to troubles than chicory, but do need to be protected in winter.

Nutritional value

It is a pity that it is not eaten more, as it has a comparatively high iron content, a great deal of carotene and useful amounts of vitamin C and dietary fibre; mineral salt and protein content are good. Together with chicory, perpetual and winter spinach, Chinese cabbage and the brassicas there should be no shortage of fresh green salading in winter to boost one's mineral intake at a time of year when these are most needed and most likely to be in short supply.

Culinary uses

The use of the endives in salads is endless; their flavour adds sharpness to the blander ingredients and suggests the use of mayonnaise rather than vinaigrette as a dressing. They are seldom cooked, but can be treated like spinach, blanching them first.

Cultivation

Cultivation is not difficult; the main point to watch is to prevent the curly-leaved kinds bolting in summer, and to prevent rotting in late autumn and winter. Slugs and snails are not half as fond of them as of lettuce and chicory. The most useful sowings are made in summer.

CULTIVATION

Site and soil sunny/open; rich, fertile, well-drained soil, not clay
Soil preparation dig 1 spade deep in autumn and mix in organic matter; fork in fertilizer 2 weeks before sowing
Sow outdoors; curly, mid-spring–late summer, for summer and up to mid-autumn use; Batavian midsummer–early autumn, cover later sowings, for autumn–winter cropping; sow thinly 1.5cm (½in) deep and space rows 30cm (1ft) apart
Thin curly, to 23cm (9in) apart, Batavian to 30cm (1ft)
Care keep well watered, especially seedlings, curly vars. and in drought; keep free of weeds; cover with cloches from early autumn
Blanch curly kinds partially with inverted plate placed over leaves; Batavian, tie leaves together and cover with inverted pot, cover drainage hole; if in pots in shed, use pots or straw and keep dark; both kinds must be dry before blanching
Harvest cut heads close to ground, 3 weeks after starting to blanch summer–autumn, 5 weeks winter
Troubles rotting, lack of protection, wet when blanched, slug/pest injury; slugs/snails; bolting

RECIPE
Exotic

Caesar's salad
1 curly endive, blanched
1 garlic clove
50g/2oz blue cheese
vinaigrette of oil, lemon juice, salt, black pepper, Tabasco sauce
fried croûtons
Separate the endive leaves, wash and pat dry; make the vinaigrette using a dash of Tabasco and 2 parts oil to 1 of lemon juice, then mix the crumbled blue cheese into it until thoroughly blended. Wipe the peeled, cut garlic clove over the inside of the serving bowl, put the endive leaves in it and pour the dressing over them, then toss until all are coated. Add the croûtons and garnish with snipped chives and slices of lemon, if desired.

ESSENTIAL FACTS

Type of plant annual or biennial
Part eaten leaves
In season midsummer–early spring
Yield 9–13 plants/3m (10ft)
Time from sowing to harvest 12 weeks
Size 10–15 × 20–30cm (4–6 × 8–12in)
Hardiness hardy
Seed viable 5 years
Germination period 5–21 days

CULTIVATION

Site and soil sheltered, sunny; fertile, well-drained, preferably sandy

Soil preparation dig 1 spade deep in winter and mix in organic matter

Sow outdoors, midsummer 6mm (¼in) deep, in rows 30cm (1ft) apart; can be sown in mid-spring in sheltered mild gardens and warm temperate climates

Thin to 20cm (8in) apart, thin early, otherwise may not bulb; thin carefully to disturb retained seedlings as little as possible

Care never allow to run short of water; watch for slugs/snails; mulch, use black plastic which keeps soil warm as well as moist; liquid-feed once when bulbing begins; earth up in stages to blanch, from time when bulb is size of hen's egg

Harvest cut bulb off just above ground-level when it is about 7–10cm (3–4in) diameter

Troubles no bulb, lack of water, not enough nutrient, cool conditions, lack of sun, careless thinning; bolting, dry soil, sowing too early in cool temperate climates; slugs/snails when plants young

Fennel

(Finnochio *Foeniculum vulgare* 'Dulce') UMBELLIFERAE

The herb fennel has a long and ancient history, since it was described in Egyptian writings of 1500 B.C., and it was a common plant in the Greece of classical times, when it was known by its Greek name of *marathon*, from the district in which it grew in large quantities. The vegetable or Florence fennel has a similarly long history, during which it became particularly popular in Italy; both sorts were eaten for the seeds as well, and were considered to have medicinal value. Two sorts of fennel, wild and sweet, were described by a leading British herbalist in 1596. He recommended the powdered seed as being good for the eyesight.

Varieties

There are two sorts of fennel: the herb, called *Foeniculum vulgare* and sometimes known as wild fennel, and the vegetable which, besides the common name given above, is also referred to as sweet fennel. Florence or sweet fennel is cultivated for the swollen, bulbous base to the stems, which interleave and fold over one another to sheathe the central growing point from which eventually comes a 75cm (2ft) stem topped by yellow flowers in summer. It has feathery needle leaves like those of its herbal relation.

The flavour is distinctly of aniseed, and the bulb slices into crisp segments which add a crunchy texture to salads in the same way that celery does, but with a completely different taste.

Nutritional value

Fennel has little nutritional value, being low on protein, and vitamins; the calorie count is also low, but there is a useful quantity of dietary fibre, and a great deal of water. The flavour is provided by a volatile oil called anethole, found throughout the plant, and it is this which makes it such an appetizing vegetable, especially when thinly sliced and eaten raw. It has in fact medicinal qualities, as a stimulant to appetite.

Culinary uses

Steaming or braising are the best ways of retaining the flavour, and it can be served with melted butter, cheese sauce, or baked in the oven. It makes an excellent accompaniment to fish, and can provide a main course for vegetarian meals.

Cultivation

Fennel is essentially a late summer and autumn vegetable, and in cool temperate climates it can be difficult to get it to bulb unless the summer is a good one. It needs a sunny sheltered position, not shaded by other vegetables and also quite a lot of moisture to prevent it running to seed. If protected from the beginning of mid-autumn, the bulbs will often stand in good condition until late autumn, if not later.

ESSENTIAL FACTS

Type of plant biennial
Part eaten bulbous stem bases,
leaves
In season late summer–mid-
autumn; warm climates, early
summer–late autumn
Yield 16/3m (10ft); bulb weighs
125g–250g (¼–½lb)
Time from sowing to harvest 8–10
weeks
Hardiness need winter protection
Seed viable 3 years
Germination period 6–12 days

RECIPE
Exotic

Fennel sauce (to serve with fish)
500g/1lb fennel
50g/2oz butter
1 clove garlic, crushed
1 tbls chervil or parsley
double cream, to dilute
½tsp Pernod (optional)
Remove the coarse outer leaves,
then boil the fennel bulbs whole in
boiling salted water for about 15–20
minutes. Drain, then stew over low
heat in butter for about 10 minutes,
until soft. Purée in a blender or food
processor. Reheat gently if
necessary and add 1 or 2tbls double
cream, and Pernod if desired.
Sprinkle with chopped fennel
leaves, to serve.

Leek

(Allium ampeloprasum 'Porrum'*)* LILIACEAE

The mild onion-like flavour of leeks has made them popular as an edible plant for many thousands of years; their use has been recorded in Sumerian writings, in Egypt during the Pharaohs, and in the Bible. The Romans knew and valued them such that Nero ate leeks in considerable quantities, in oil, to clear his throat and restore his voice for perorations. During mediaeval times the leek was a popular vegetable in France, particularly near Arras, and is one of the vegetables listed by Jon the Gardener, in the first book about gardening written in English.

For hardiness, the average leek would be a difficult vegetable to better. It will stand through the coldest winter, start to grow again in spring and be fit for eating even in late spring, before starting to run up to flower in summer. Many late plantings of leeks, planted in late summer or early autumn as a last resort, have put on sufficient weight as soon as the winter was over, to fill the 'hungry gap' when vegetables are at a premium, and so justify planting the pathetic specimens of a previous year.

Nutritional value

Nutritionally, they have good food values, with plenty of mineral salts including iron, a good deal of vitamin C and carotene, dietary fibre, protein, and not too many calories. Incidentally the leaves contain an enormous quantity of carotene, so the green part should be eaten with caution, as too much vitamin A can be harmful. With this high food value, it can be seen that no self-respecting vegetable garden should be without its complement of leeks, especially as they are in season at a time when fresh vegetables are at a premium.

Culinary uses

While boiling leeks whole is the standard method of cooking, steaming is preferable for retaining the nutrient value, and cutting them transversely into rings will make for quicker cooking and avoid the tendency to sliminess that occurs with over-cooking. Apart from use as a side-vegetable, leeks are an excellent constituent of soups, casseroles and meat mixtures for pies. In France, they are frequently served braised in wine. They are also suitable for salads and as starters, cooked and then served cold.

Cultivation

Leeks are grown in cool and warm temperate climates throughout the world. One of the easiest vegetables to grow successfully, provided they are started early enough, they need little care, and have few troubles. The biggest and most succulent leeks need a rich moist soil and, although they can be blanched by earthing up, better results are obtained by wrapping them, as celery is.

CULTIVATION

Site and soil sun/open; most soils, good drainage preferred

Soil preparation dig 1 spade deep in winter and mix in organic matter in generous quantities; rake compound fertilizer into light soil 1 week before planting, and tread to firm it

Sow seed thinly outdoors 6mm (¼in) deep in nursery bed early–mid-spring, or in early summer for late crop

Thin carefully to 4cm (1½in) apart final spacing

Plant transplant when about 20cm (8in) tall; trim back roots and leaf tips; space 10–15cm (4–6in) apart, 30cm (1ft) between rows; plant so that the leaves arch along the row, not across; water day before if soil dry; make hole 15cm (6in) deep with dibber, drop plant into it and water in gently with spout to settle roots, do not fill hole with soil

Care keep well watered; allow soil to fill hole naturally so blanching stem; when full, draw dry soil round stems occasionally to increase length of blanched stem until end of mid-autumn; liquid-feed until the end of late summer for bigger leeks (but flavour loss if too big)

Harvest dig as required from early autumn, starting before fully grown; remove alternately so that remainder can enlarge

Problems virtually none; occasionally small caterpillars inside leaves

ESSENTIAL FACTS

Type of plant biennial
Part eaten stem, leaf
In season early autumn–late spring
Yield 125–250g (¼–½lb)/plant
Time from sowing to harvest 28
weeks, 40 if late var. or sown late
Size 30 × 30cm (1 × 1ft); blanched
stem average 15cm (6in) long
Hardiness hardy
Seed viable 4 years
Germination 9–21 days

RECIPE
Healthy

Leeks à la Grecque
4 leeks
50g/2oz brown rice
12 black olives
1tbls tomato paste
salt and black pepper
150ml/¼pt oil, preferably olive
300ml/½pt water
chopped parsley
slices of lemon
lemon juice
Cut the leeks into 4cm (1½in)
lengths; put them together with the
tomato paste, oil, water and
seasonings into a pan, bring to the
boil, and simmer with the lid on for 5
minutes. Then add the rice, bring to
the boil and simmer again, still
covered, for another 8–10 minutes.
Take off the heat but leave for about
15 minutes, when the rice should be
still firm though cooked. Mix in
lemon juice to taste, chill and serve
decorated with olives, slices of
lemon and parsley.

Lettuce

(Lactuca sativa) COMPOSITAE

The lettuce has always been a popular vegetable. Its use was noted in the records of the Sumerian civilisation and the Greeks and Romans used it for medicinal purposes as well as culinary. Indeed, it is not long since its use as a sedative ceased, following the breeding-out of the bitter flavour in the original species. Another species, *L. virosa*, was specifically grown to provide an ingredient used as a cheap substitute for opium during the 18th and 19th centuries in western Europe.

Varieties

The number of lettuce varieties now available is bewildering, the more so because the groups into which they fall are specific to certain seasons of the year. It is quite easy to cut lettuce all year round, with the help of greenhouse protection, though for those without this facility, the red chicories can be substituted, but it is important to choose the right variety for the right reason.

Besides this grouping by seasons, lettuce can also be divided according to their shape; there are roughly four categories, with some overlapping. The most popular and well known is the cabbage or butterhead lettuce, round with a firm central heart, and rounded, soft leaves; the crispheads are nearly as popular. These are much larger lettuces with a solid tightly packed, chunky heart, crunchy stems and crisp leaves with fringed edges, often called the 'Iceberg' varieties. Also a rounded lettuce, but open-growing and flat-hearted are the cut-and-come again type, such as the 'Salad Bowl', from which single leaves are stripped as required, each leaf being wavy-margined and long. These can be cropped through the summer. A fourth group are the cos lettuces, upright growing with long, dark green leaves, forming a large plant. Intermediate between this and the cabbage lettuces are a few varieties which are smaller, with a definite heart, but still upright, and with a particularly nutty flavour – they are extremely hardy and do well if protected by cloches through the winter.

Nutritional value

Nutritionally, lettuce can provide raw green vegetable material all year, which contains a great deal more carotene than most vegetables, together with vitamin C. Lettuce also

CULTIVATION

Site and soil sunny/little shade; medium–well broken down heavy soil, add lime if acid

Soil preparation dig 1 spade deep in winter and mix in rotted organic matter; add general compound fertilizer 7–10 days before sowing if soil light

Sow outdoors early–mid spring under cloches for early summer crop; sow mid-spring–midsummer for midsummer–mid autumn crop, cloching in early autumn; sow late summer for mid–late autumn crop in mild gardens; sow early–mid autumn and cloche at once for early–late spring cropping sow 3–6mm (⅛–¼in) deep in rows 20–30cm (8–12in) apart

Thin to 15–30cm (6–12in) apart, but thin late summer–autumn-sown kinds to about 7cm (3in) and thin again late in the winter to final spacing

Care keep weeds under control especially in early stages; watch for slugs/snails; keep well watered in dry weather; liquid-feed on light soils in summer

Harvest cut cabbage types when hearts firm and just before they start to unfold before bolting; cut cos when maximum height reached; harvest by pulling the root up; if the lettuce is to be kept, wash the root and store complete in polythene in the refrigerator; harvest leaf lettuce by taking single leaves as required

Troubles slugs/snails on young plants and seedlings; protect from birds when young; greenfly; root-aphids; soil-living caterpillars; grey mould; mildew

Iceberg

All year round

ESSENTIAL FACTS

Type of plant annual
Part eaten leaves
In season all year
Yield 10–15 heads/3m (10ft) row
Time from sowing to harvest 8–12 weeks
Size cabbage, 15–25cm (6–10in) square; cos, 20–30 × 10–12cm (8–12 × 4–5in)
Seed viable 3–4 years (3 for cos)
Germination period 10–18 days

Avoncrisp

All year round

Cos

contains dietary fibre, mineral salts, few calories, and a large amount of water, so much in fact that it must be a great thirst quencher, hence its summer popularity.

Culinary uses

In cooking its first use must always be as a salad ingredient, but it can be braised, made into soup or used as a delicate casing for stuffings. In salads, red-tinted lettuce, and the white-green centres of the crispheads provide colour alternatives, and if garnished with flowers such as borage and marigold will present a more appetising and interesting appearance.

Cultivation

The cultivation of lettuce is now down to such a fine art that it is almost a case of choosing the right variety for the right month, but unless there is a lot of space to spare, it is not worth attempting to grow more than two or three different kinds depending on preference. However, for winter production, it is essential to choose the short-day varieties of cabbage lettuce, i.e. those that will heart up without much light. Apart from this seasonal requirement, good lettuce require a moist, well-cultivated, fertile soil and plenty of water throughout their life. Lack of this in summer results in bolting – running a stem up to flower and seed – and also attack by root-living pests.

RECIPE
Healthy

Greek salad

1 lettuce (crisphead or 'Winter Density' cos), washed
½ cucumber, sliced thinly
1 onion, sliced thinly
4 tomatoes, quartered
12 black olives
1 sweet red pepper, thinly sliced
vinaigrette made with olive oil
garlic, chopped, to taste
125g/4oz Fetta cheese, cut into four fingers

Tear the lettuce leaves, and mix them with the remaining ingredients except the cheese, and then toss all with vinaigrette. Wipe the cut surface of a garlic clove round the inside of the salad bowl, then pile in the salad and top with the fingers of Fetta.

Marrow & Courgette

(Vegetable squash, marrow squash; zucchini, baby marrow
Cucurbita pepo) CUCURBITACEAE

The plebian marrow has suddenly become popular during
the last decade, since baby marrows are what are known as
courgettes or zucchinis. Although now common in cool
temperate climates, they have always been an essential
vegetable in Mediterranean cuisines, notably Italy. When
home-grown, they become a delicacy. In fact, if cut when
they are literally only 7–9cm (3–3½in) long and cooked at
once, they have a delicious, fleeting flavour not present in the
adult marrow, and indeed not present when they are more
than 2 hours picked. Almost more than any other vegetable,
they pay for being grown in the garden to ensure that
tantalisingly gourmet flavour. Mature marrows, with their
coarser flavour have become less popular.

Nutritional value

The main content of the marrow is water, only exceeded in
this by the cucumber, and when cooked it outdoes even that
in its liquidity, so that it is fortunate it is a hot-weather
vegetable. Nevertheless, it does contain some minerals and
vitamins as well, so ground reserved for cultivating it is not
wasted, especially as the green or golden courgettes make
good salad ingredients, whether cooked or raw, and the
flowers are decorative enough to warrant it a place in the
ornamental vegetable garden.

Culinary uses

Marrows and courgettes have many culinary uses; as
accompanying vegetables; marrows as a main course stuffed
with various savoury mixtures, in chutneys, for wine and as
jam; courgettes in salads, as part of ratatouille, or dipped in
batter and fried. The flowers are also edible, and in
Mediterranean countries are often fried, coated with batter,
or stuffed with a meat filling.

Cultivation

In spite of their tropical origin, marrows are easily grown in
cool temperate climates during the summer, although they
have to be germinated under cover, and may need hand-
pollinating in dull weather. The female flowers have a tiny
oblong swelling between them and the stalk. The most
southern parts of America, together with Central America,
are probably their original home, from which they were
widely distributed by the Indians before the European
settlers arrived. They crop profusely, and using the tiny
marrows is a convenient way of keeping up with them,
otherwise it is common to find that there are three enormous
marrows lurking under the leaves when, practically the day
before, there were none there at all.

CULTIVATION

Site and soil sunny or open, most
soils, preferably moisture-retentive
but not waterlogged

Soil preparation dig 1 spade deep
in early spring, and mix in organic
matter; fork in fertilizer 2 weeks
before planting, warm soil with
cloches in advance

Sow in mid–late spring singly in 7cm
(3in) pots, 18°C (65°F); sow on edge
2.5cm (1in) deep

Pot on into 10cm (4in) pots and
maintain temperature until few days
before planting, then harden off

Plant early summer outdoors or
whenever frost unlikely; space bush
varieties 75cm (2½ft) apart each
way, trailing kinds 120 × 60cm (4 ×
2ft); plant so that each is on a shallow
mound; water in

Care make sure that water does not
collect round base of stem in soil
depression; stop main stem of
trailing vars. at 150cm (5ft); water
heavily in dry weather; mulch;
liquid-feed from early in late
summer to early autumn; hand-
pollinate in dull, cool weather

Harvest cut courgette/zucchini
when about 9cm (3½in) long; cut
marrows at 30–38cm (12–15in) long,
before lifting; keep cutting to ensure
continuity of production

Store allow last marrows to reach
maximum size or cut last marrows
just before frost, and put in nets in
frost-free dark place – they will
keep at least 2 months

Troubles generally healthy; slugs/
snails; collar rot, stem rots at soil
level, due to waterlogging; small
embryo marrows wither at tip due
to: irregular water supplies, bad
drainage, too much pruning, too
much water; no fruit, lack of water,
or no pollination

Type of plant annual
Part eaten fruit, flower
In season midsummer–mid-autumn
Yield depends on size at which cut:
courgette 15–20/plant 56-84g (2–3oz)
each; marrow 5–6/plant ½–2kg (1–
4lb) each
Time from sowing to harvest 12–14
weeks
Size fruit, 9–38cm (3½–15in); plants,
bush 60–90cm (2–3ft), trailing 60 ×
180cm (2 × 6ft)
Hardiness tender
Seed viable 7 years
Germination period 6–12 days

RECIPE
Exotic

Courgette paté
1kg/2lb courgettes
1tbls salt
50g/2oz butter
4 eggs
300ml/10fl oz soured cream
1tbls chopped parsley
1tbls chopped fresh basil
pinch cayenne pepper
freshly ground black pepper
Garnish (optional)
3tbls thick cream, whipped
finely chopped parsley and basil
Coarsely grate courgettes in a food
processor. Place in a colander,
sprinkle with salt and leave to stand
for about ½ hour; then rinse
thoroughly under cold running
water and dry with paper towels.
Melt butter in saucepan, add
courgettes and cook slowly for
about 10 minutes until soft. Set aside
to cool. Meanwhile, heat the oven to
180°C/350°F/gas mark 4. Line a 1.5l/
3pt casserole with buttered
greaseproof paper. Beat together
eggs and cream. Add courgettes,
herbs and seasoning and stir well
until evenly blended. Pour mixture
into prepared tin, cover with foil and
cook in a pan of cold water in the
oven for about 1 hour or until paté is
firm. Leave to cool in the tin, then
turn out onto a serving dish. Spread
whipped cream over the top of the
paté and sprinkle with the finely
chopped herbs. Serve as a starter
with whole cooked prawns.

Mushroom

(Agaricus campestris, Psalliota hortensis) AGARICACEAE

Mushrooms have been highly valued for their flavour for thousands of years and have also been regarded with some awe, since the ancient Egyptians thought they contained magical properties, and the Chinese valued them for medicinal powers. The Romans knew and loved them, and because they grow wild they have been eaten throughout Europe including Britain, certainly since Neolithic times. (For other edible fungi, such as truffles and cep, see Unusual and Exotic Vegetables, pages 132–137.) However, their formal cultivation as a crop is quite recent; only at the beginning of this century did it become possible to produce spawn with certainty from mushroom tissue; after that mushrooms could be guaranteed, and selected for a particular characteristic.

Nutritional value

Nutritionally, mushrooms are surprisingly valuable; they contain a variety of minerals, a comparatively high protein percentage, vitamin C and other vitamins in lesser amounts, and dietary fibre, together with a low calorie count. Home-grown and field mushrooms are said to contain quite large quantities of protein, but no figures are available.

Culinary uses

As with tomatoes and onions, the mushroom is a versatile vegetable for culinary purposes; there is almost no savoury dish to which it cannot be added, and eaten alone, the delicate flavour takes a good deal of beating. It can be fried or grilled, added to all kinds of meat stews and casseroles – it is an essential ingredient of coq au vin – made into soup, stuffed, added to salads cooked or raw and included in

CULTIVATION

Site and soil under cover, shaded, even warmth of about 16–18°C (60–65°F), ventilated but not windy, some humidity; rotted stable manure or rotted wheat straw

Soil preparation allow fresh stable manure to rot down in a heap 150 × 150 × 150cm (5 × 5 × 5ft) on a stone or concrete base; smaller than this will not give satisfactory results, or use straw and special mushroom activator, at the rate directed by the manufacturer. Water well and cover the top with sacking or plastic sheet; when the heat generated reaches 71°C (160°F) turn, and repeat turning weekly, and watering if at all dry, until heap is crumbly, dark brown and sweet-smelling. Alternatively, obtain stable manure already in this state from suitable source, e.g. farmer, riding stables. Fill containers at least 23cm (9in) deep, not more than 30cm (1ft), firm down and leave until temperature is 24°C (75°F)

Spawn use proprietary spawn, either block or grain, and push into compost 2.5cm (1in) deep; keep at 16–18°C (60–65°F) and do not allow to dry out

Case when grey threads appear on surface, cover with moist casing, a mixture of 2 peat, 1 chalk, parts by volume, 5cm (2in) deep

Care maintain even temperature; keep casing moist but not sodden by spraying with water every 2–3 days, but be guided by state of casing rather than a rigorous timetable

Harvest pick about 4 weeks after casing; mushrooms will develop in flushes, producing about 4 flushes in 2 months, but this varies considerably, depending on correctness of watering and maintenance of temperature, and mushroom development may be continuous but in smaller amounts at a time; pick by twisting off the stem close to the compost and fill in hole with casing; use compost in garden when crop has finished

Troubles mushroom fly, the maggots of which feed on the developing mushrooms, use insecticide as soon as flies seen

omelettes, quiches, pizzas and risottos. If eaten raw, it is best to use the buttons; they make a good starter with a suitable dip. The best ones to cook and eat in their own right are the large ones which are fully open, known as 'flats'; the 'caps' are the inbetweens. Varieties of dried mushrooms are now increasingly available and are widely used in Oriental cuisines.

Cultivation

Mushrooms are a unique crop in that they do not need sunlight to grow; they will grow whether it is light or dark, but it is better to shade them during the day, as strong sun can discolour them and makes it difficult to maintain the even temperature which is so important. A steady supply of moisture is also vital, but it has to be even; too much or too little will stop production. Given these conditions, however, mushrooms can be grown all year round.

They are saprophytes, that is, they live on dead organic matter, hence the reason for using composted animal manure or straw as the growing medium. This has to be thoroughly rotted first to ensure that organisms that could destroy the mushrooms are themselves eradicated in advance. On top of the compost a layer of material is placed, known as the casing, which helps to prevent the surface of the compost from drying out. Mushrooms can be grown at home on home-made compost, which can be inconvenient because of the amount required for successful composting, or a proprietary pack can be bought, in a bucket or box, containing compost ready spawned, together with casing material. Such packs come with instructions and give good results, but are of course more costly.

ESSENTIAL FACTS
Type of plant annual, without green colouring, saprophyte
Part eaten fruit
In season all year
Yield 750–1000g (1½–2lb)/sq ft depending on suitability of compost
Time from spawning to harvest about 5–6 weeks
Size 2.5–10cm (1–4in) diameter, mushrooms, height 4–7cm (1½–3in)
Hardiness hardy
Spawn viable dry spawn 6 months, moist spawn 5 days
Spawning period 10–14 days

RECIPE
Exotic

Champignons à la grecque
3 tbls olive oil
3 tbls red wine
2 medium tomatoes, skinned and
    chopped
salt to taste
250g/8oz button mushrooms
3–4 crushed peppercorns, crushed
6 coriander seeds, crushed
sprig of thyme
1 bay leaf
Put all the ingredients but the mushrooms into a small saucepan and bring gently to the boil, then simmer 2–3 minutes. Wipe mushrooms clean with a damp cloth, then sprinkle with a little lemon juice and rub in. Add to the pan and simmer 5 minutes. Take out the mushrooms and put in a shallow serving dish; continue cooking sauce briskly 2–3 minutes or till thick and reduced. Pour over mushrooms and serve cold.

91

Okra

(Gumbo, Ladies' fingers, *Hibiscus esculentus*) MALVACAEAE

Okra is a tropical plant originally only found growing wild in Africa, but which has now spread to the West Indies, Asia and tropical North America. It is related to the cotton plant – they are both in the same plant family, the *Malvacaeae* – and grows to about 90–120cm (3–4ft) in cultivation, 180cm (6ft) and more in the wild. Its large, softly-lobed leaves and open yellow flowers with a crimson centre, make it a sufficiently decorative plant to grow in the ornamental vegetable garden.

The seed-pods which follow the flowers are the part eaten; they are grooved, finger-shaped pods, mostly best harvested when about 7cm (3in) long, though some varieties are still succulent when 20cm (8in) long. They should still be soft, but snap crisply in half when broken. If buying them, reject any which are shrivelled and a poor colour, they will have lost their flavour and be tough and stringy.

Culinary uses

Okra pods can be eaten raw, as well as cooked, and can also be dried. They are relished for their flavour, and their mucilaginous texture, which makes them a popular ingredient for thickening soups. Chicken gumbo is a traditional Creole dish of the southern states of North America which includes ham and onions as well as the chicken and gumbo of the title, and the pods are much used in vegetable curries, fritters, or casseroles, and steamed and served as a side vegetable.

Nutritional value

The raw pods have a surprising amount of protein in them, and there are also quantities of vitamin C and A as carotene, as well as a range of minerals, of which potassium has the highest content. The calorie value is low, and there is a useful amount of dietary fibre.

Cultivation

Cultivation is not difficult in tropical climates; the plants grow rapidly from seed, and pods can be harvested within two months. Cropping will continue for a further eight weeks or so. In warm temperate climates artificial warmth will be needed to start the seed and while the plants are young, and in cool temperate regions, it is necessary to keep the plants in a greenhouse and grow them like tomatoes or aubergines unless the summer is unusually hot or the garden is particularly sunny and sheltered during the growing season.

The plants are affected by the same pests and diseases as cotton plants are, and in some countries both species have a ban on their cultivation at a certain time of the year in an effort to prevent an epidemic of these problems. For the average family, half a dozen plants are ample.

CULTIVATION

Site and soil sunny, sheltered; rich moist, deep soil or fertile well-drained compost such as J.I. No 3

Preparation dig at least 1 spade's depth in early spring, and mix in rotted organic matter; rake in a light dressing of a potash-high compound fertiliser a week or so before sowing or planting

Sow tropical: sow outdoors in rows spaced 60cm (2ft) apart; sow thinly 1.5cm (½in) deep; temperate: sow two or three in 5cm (2in) pots early–late spring in 18–24°C (65–75°F)

Thin outdoors: when large enough to handle to a spacing of 45cm (1½ft); in pots: to the strongest in each pot

Pot cool-temperate: pot on successively as roots fill pots until 25–30cm (10–12in) size reached, using J.I. potting compost No 3 for final size, or plant in greenhouse border when plants 10cm (4in) tall in late spring, space 30cm (1ft) apart, and provide heat at night, and in day if necessary to ensure 21°C (70°F) during daylight

Plant warm temperate: plant outdoors when all risk of frost is past, at the spacings specified above

Care keep weeds under control; watch for slugs and snails when plants young; water well in dry weather

Prune/support supply canes or other supports for plants in windy sites and in the greenhouse, especially for tall varieties

Harvest harvest the pods while still immature, usually about 4–5 days after the flowers open, when the seeds are still only half-formed and the pods soft and tender; keep harvesting to ensure a succession, and cut rather than pick to avoid pulling the plant out of the ground

Troubles slugs and snails; greenfly; leaf-eating caterpillars; red spider mite; weevils; grey mould; mildew

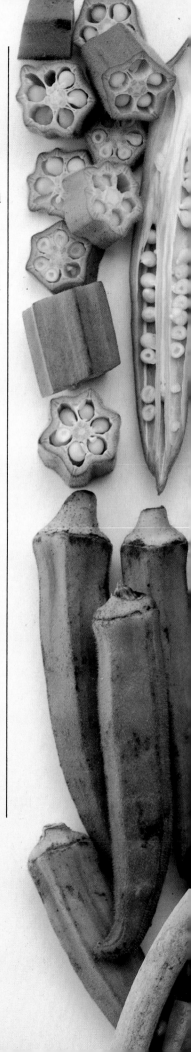

Type of plant annual
Part eaten seed-pods
In season midsummer–early
autumn
Yield 15–20 pods/plant
Time from sowing to harvest
16 weeks
Size 90–180cm (3–6ft) tall; pods
7–20cm (3–8in) long
Hardiness tender
Seed viable 4 years
Germination period 5–9 days

RECIPE
Economical

Corn and tomato gumbo
1tbs oil
175g/4oz onions, chopped
175g/4oz streaky bacon
*1 red or green pepper, seeded and
    chopped*
*250g/¹/₂lb tomatoes, peeled and
    chopped*
*250g–300g/8–10oz sweetcorn
    kernels*
500g/1lb okra, trimmed
1tsp sugar
1 dried chilli, chopped
salt and pepper, to taste
Melt the oil in a heatproof casserole,
add the onions and bacon and cook
gently until the onions are beginning
to soften. Add the tomatoes and
cook for a couple of minutes to
release their juice, then add the
remaining ingredients. Cover the
casserole and simmer for about 25
minutes, stirring occasionally, until
the vegetables are tender.

CULTIVATION

Site and soil sunny; well-drained, most soils, not acid, manured for previous crop

Soil preparation dig 1 spade deep in autumn; lime if acid; rake in fertilizer 2 weeks before sowing/planting; clean off weeds thoroughly if sowing

Sow outdoors early–mid-spring, soil should be workable, not cold or wet; sow 0.6cm (¼in) deep, rows 23cm (9in) apart; Japanese var., sow about the middle of late summer, but time of sowing depends on area, earlier in cold areas, later in warm ones, sow seed 2.5cm (1in) apart, rows 23cm (9in) apart

Thin in 2 stages, to 2.5–5cm (1–2in) when seedlings straight, then to 7–20cm (3–8in), depending on size of bulb wanted; Japanese var., thin in spring to 10cm (4in); spring onions, thin to 2.5cm (1in)

Plant sets outdoors early–mid-spring, not in cold wet soil; plant so that tip just showing, cut off long trailing withered stalks; space as above

Care keep well weeded in early stages; push any displaced sets back into soil during 1st weeks after planting until well-rooted; water if dry weather prolonged; liquid-feed occasionally; remove any flowerheads as soon as they appear

Harvest seed-grown, ready in early autumn; set-grown, in late summer; Japanese, in early summer; spring, in late spring–midsummer, depending on sowing date; dig up bulb onions 7–10 days after the leaves start to yellow and wither, spread out singly to dry, in sun outdoors in good weather, indoors if wet in warm dry place, for about 7–10 days; store in cool, light place, hang in bundles by withered stems

Troubles few, but can be bad; onion fly, wilting yellow leaves, plant ceases to grow, maggots in bulb, destroy and treat soil next year with soil insecticide before sowing/planting; white rot, wilting yellow leaves, soft bulb, white mould on base, no maggots, destroy and do not grow onions there for at least 8 years; bullneck, a thick shank and small bulb, over-manuring or too much nitrogen in liquid-feed, or sowing too deeply; bolting, too hot and too dry, loose soil at planting/sowing time, cold and/or wet at planting/sowing, poorly treated sets

Onion

(Allium cepa) ALLIACEAE

The onion is one of the longest and most widely used vegetables, with a recorded history reaching back to 3,500 B.C. Where the wild onion originated is not definitely known but it is thought that its native habitat is somewhere in Central Asia. The ancient Egyptians thought so highly of them that they were used as offerings to their gods, and one sort even deified. Enormous quantities of them were eaten; they were used in medicine and in mummification, and the slaves building the pyramids ate onions, as well as garlic and radishes.

The Romans gave them our name of onion, which is derived from the Latin word *unionem* or *unio*, meaning single, referring to the single bulb it produces, unlike other varieties of the same species. Nero ate them to cure colds, and the Roman belief in their medicinal use ran alongside the liking for the flavour and they were used for sleeplessness, coughs, sore throats and stomach upsets.

Varieties

Onions are tremendously varied in shape and size, though this is seldom appreciated until they become home-grown rather than shop-bought. Bulb onions tend to be 'flat', that is, a flattened globe, or plump and round, with light brown skins, and are usually fairly strong-tasting. Much larger onions, called Spanish onions, though not necessarily coming from Spain, have a mild flavour and are easily eaten raw so are good for salads; they are pale yellow-, white- or red-skinned, with a reddish tinge to the flesh in the last-named. Bulb onions are the kind which are ready in late summer and store through the winter until mid spring.

A recently produced variation, the Japanese onion, on the small side and strongly flavoured, will mature in early summer and so bridge the gap, but is not suitable for storing. Spring onions (scallions) used in salads are onions grown from seed sown thickly, and the thinnings used when they have grown to a suitable size. Pickling onions are a special variety, producing small, white round bulbs which never grow large, and shallots are a variety of onion, the bulb of which simply splits into separate parts, up to 10 in a well-grown plant.

The Egyptian or tree onion, *A. cepa* 'Aggregatum', is of Canadian origin, and has stout hollow stems, up to 90–120cm (3–4ft) tall, with clusters of small bulbs at the ends, occasionally mixed with flowers; these can be harvested in early autumn.

Continued on page 96

Brunswick

Quicksilver

Sturon

Shallot

ESSENTIAL FACTS

Type of plant biennial
Part eaten bulb
In season late summer–mid-spring; Japanese, early summer–late autumn; spring, late spring–midsummer
Yield 1 bulb average 125g (4oz), can be from 60–180g (2–6oz); 100 sets can give 11kg (24½lb)
Time from sowing/planting to harvest seed, 6–10 months; sets, 4 months
Size bulb, average 5–7.5cm (2–3in) wide; leaves 45cm (1½ft) tall
Hardiness hardy, but overwintered need protection in very wet/cold weather
Seed viable 1–2 years
Germination period 21 days

The Welsh, or Japanese bunching onion, also called Ciboule (chibol, sprout), *A. fistulosum*, also consists of clusters of bulbs with brown skins, rather elongated, which mature early in the year and can be treated like spring onions.

Nutritional value

Onions contain few calories, an appreciable amount of calcium and some vitamin C but their nutritional value is not high, despite their reputed medicinal qualities.

Culinary uses

Onions are possibly the most widely used vegetable in the world. No casserole or stew is complete without them and they are added to thousands of other dishes as a moisturizer and flavouring. On their own, they can be baked or boiled, or made into a delicious sauce or soup.

Cultivation

Probably the major reason for the worldwide distribution of the onion, apart from its attractive flavour, was the ease with which it can be carried, and its long dormancy period. All bulbs are plants with instant packaging, formed to survive long, hot dry summers in most cases, or long periods without rain, sometimes associated with cold, and the onion is one with considerable potential for survival. In cultivation it is grown to produce as large a bulb as possible; the vegetative stage of its life is the important one, and flowering has to be avoided. Since it is biennial, this should be easy, but it can bolt in too hot and dry conditions, and then the bulb does not swell, and one is left with a large and handsome, round seedhead.

Ailsa Craig

Sets of sturon

White Lisbon

RECIPE
Economical

French onion soup

4 large onions, finely chopped
50g/2oz butter
3 tsp flour
1 litre/2 pints stock or water
French bread cut into thick slices
125g/4oz Gruyere or Parmesan
  cheese, grated

Melt the butter in a saucepan, add the onions and cook very gently until the onions are soft and highly coloured. Add the flour increase the heat a little and cook, stirring constantly until the onion has turned as brown as it can without burning (this may take about 20–30 minutes). Bring the stock or water to boiling point, pour over the onions and simmer gently for about 15–20 minutes. Ladle the soup into individual heatproof bowls and top each bowl with a slice of French bread. Sprinkle grated cheese over the bread. Put the bowls under a hot grill until the cheese begins to melt and bubble, then serve immediately.

Sturon

CULTIVATION

Site and soil sun/little shade; deep, moist, fertile, alkaline, free of large stones
Soil preparation dig 2 spades deep in late autumn–winter, mix in lime if soil acid; use ground manured for previous crop
Sow outdoors, early–mid-spring 1.5cm (½in) deep; cover with cloches if early sowing; station sow 15cm (6in) apart 3 or 4 at each station; rows 20–30cm (8–12in) apart depending on var.
Thin to 1 at each station
Care keep free of weeds while seedlings; mulch after thinning; water in prolonged drought
Harvest dig up from mid-autumn onwards as required; mark rows if left in ground over winter; protect soil from frost for easier digging
Troubles brown or black 'shoulders' to roots, parsnip canker, due to injury (weeding, insects, cold, drought), acid soil, unrotted organic matter; cracking, irregular water supply

ESSENTIAL FACTS

Type of plant biennial
Part eaten root
In season mid-autumn–early spring
Yield 180–500g (6–16oz)/root
Time from sowing to harvest 26–30 weeks
Size root, average 12–20cm (5–8in) long, leaves 30–38cm (12–15in) long
Hardiness hardy
Seed viable 1 year
Germination period 21–28 days

Parsnip

(Pastinaca sativa) UMBELLIFERAE

The parsnip's long history of cultivation dates back to Roman times – one of the Roman Emperors, Tiberius, had consignments sent to him from an area near the Rhine, in Germany, and they were then served in a mead sauce made with honey. Parsnips had a variety of names: *parsneps*, *parsnebbs* or *pasternaks*, and were certainly known to the Saxons in Britain, as the last syllable is derived from the Anglo-Saxon word, *naep*. They always seem to have been more popular in Britain and northern Europe than in the warmer regions of southern Europe; the French have never regarded them as an essential part of their cuisine, except perhaps as an ingredient in *pot-au-feu*.

Nutritional value

They are said to contain more sugar than sugar-beet; raw parsnips certainly do have a higher sugar content than any other vegetable but this is reduced by about three-quarters, when cooked. They also have a high content of minerals, a good deal of energy value and a useful amount of vitamin C. Water content is lower than most vegetables, and there is some protein and dietary fibre present, though not much.

Culinary uses

Parsnips are unlike the majority of cooked vegetables in that they taste slightly sweet; allowance should be made for this in the foods cooked with them. The rich flavours of the traditional British roast beef and Yorkshire pudding lend themselves very well to parsnips baked underneath the joint, and pork is another meat whose taste provides a satisfactory complement. Their delicious but often undervalued flavour presents a range of possibilities when cooking them. They are sufficiently interesting to be a main course as well as a side vegetable, and to be used for jam, cakes, wine and in fritters or pancakes. They also make an excellent soup, and are particularly good when cooked in the American manner with brown sugar and butter so that they glaze beautifully.

Cultivation

Since the wild species of parsnip flourishes in meadows and by waysides in Britain and northern Europe, it is not surprising that it is easy to grow as a cultivated vegetable. Stone-free soil is important, to ensure long straight, unforked roots; the right sowing time is also another factor to watch. Tradition has it that parsnip should be sown in late winter, failing that early spring, but often snow is falling at that time. Mid-spring is usually the best time, but whatever the month, the soil should not be cold or sodden, otherwise germination is non-existent. Wait for it to dry out and to warm up; if necessary put cloches over the ground to hasten both these processes. Use new seed every year, as viability drops off rapidly after the first year. If the soil of the vegetable garden is not deep, grow one of the short-rooted varieties, which mature more quickly.

Economical

Curried parsnip soufflé

*500g/1lb parsnips, peeled and cut
  into chunks*
75g/2oz butter
75g/2oz flour
150ml/1/4pt milk
1/2–1 tsp curry powder
*1 tbsp chopped parsley or coriander
  leaves*
salt and pepper
4 eggs, separated

Boil parsnips in salted water until
tender. Drain well, reserving about
150ml/1/4 pint of cooking liquid, then
mash into a smooth purée. Grease
1.25 litre/2 pint soufflé dish with a
generous amount of butter. Make a
sauce by melting 2oz butter, adding
flour, curry powder and then milk
and reserved cooking liquid. Stir in
puréed parsnip and parsley and
season with salt and pepper to taste.
Remove from the heat and beat the
egg yolks into the mixture. Whisk
the egg whites until they form soft
peaks, then gently fold into the
mixture. Pour into the soufflé dish or
small individual dishes and cook in a
preheated oven (200°C/400°F/gas
mark 6) for about 20 minutes or until
soufflé has risen and browned on
top. Serve immediately.

Buttered parsnips When preparing
for cooking, trim and peel the
parsnips; discard the core if tough
and hard. Cut into chunks or lengths,
and boil or steam quickly until soft,
then mash with lots of butter, salt
and pepper.

Other suggestions: Partly cook,
then roast in the oven with a joint;
glaze with brown sugar; cut into
strips and deep fry, like chips, or
coat rings 15mm (1/2in) thick with
batter and then fry; mash and use to
make cakes, as potatoes are used;
cut into chunks for warming winter
casseroles; make small pancakes
with mashed parsnip, egg, salt and
pepper, plain flour and melted
butter, fry until brown.

Pea

(Pisum sativum) LEGUMINOSAE

On a prehistoric site in Burma, near the border with Thailand, pea seeds have been discovered with a carbon-dating to about 9,750 B.C.. They have been found in Bronze-age sites beside Swiss lakes in Europe and were probably brought to Britain by the Romans. Their recorded history is unbroken from then until the present day – they were referred to as 'parched pulse' in the Book of Samuel in the Old Testament.

Young peas, cooked as soon as they are picked, have a sweet succulent flavour that is completely different to those bought from a greengrocer which have had a time-lag of at least 20 hours between picking and cooking. The small-seeded peas known as petit pois are even more of a delicacy, and the mange-tout or sugar-pea type have tender pods, flat and juicy, delicious when simmered and eaten whole with melted butter.

Pea plants vary in height from dwarf to a maximum of about 150cm (5ft); they are climbing plants grown from seed, and attach themselves by tendrils which are modified leaflets, each leaf consisting of 1–3 pairs of leaflets. Their white, typical pea flowers with a keel, standard and two wings, appear from late spring to midsummer, and where there is the space to spare, peas can be picked from early summer until early autumn, though the right varieties must be chosen and sown at each season.

Varieties

The two commonest kinds are the round-seeded, which have smooth seed-coats when dried; they produce the earliest crops and can be sown in autumn to over-winter, or in early spring, and the wrinkled-seeded (marrowfat) type, whose skins crinkle on drying – they have a larger, sweeter pea,

CULTIVATION

Site and soil sunny, sheltered; deep, moist and fertile

Soil preparation dig 2 spades deep and mix well-rotted organic matter into the second spit, in early spring; add compound fertilizer 7 days before sowing or planting

Sow early spring–midsummer, protect with cloches early–mid-spring; sow mid–late autumn in mild gardens and protect with cloches, use dwarf varieties sow 5cm (2in) deep and 7.5cm (3in) apart, in 2 staggered rows 10cm (4in) apart in flat drills; allow same distance between drills as the height of the variety or sow well-spaced in seed trays, or 5cm (2in) pots, or plastic propagating packs, and plant out when large enough to handle and soil condition is right supply protection against birds; put down slug-bait or other slug deterrent

Care remove weeds; keep well-watered in dry weather when flowers opening, and as pods start to swell; mulch heavily early in plants' life; when crop finished, cut off tops and put on compost heap, dig in roots, as they will return nitrogen to the soil

Prune/train support the plants with pea sticks (bushy twigs), wires run along the row between posts, spaced 30cm (12in) apart, wire-netting, or with panels of plastic square-mesh netting strongly supported. Dwarf plants will need supports, otherwise they become flattened and the pods muddy and slug-ridden

Harvest when the pods are tightly filled with peas, or when they have stopped lengthening, for petits pois and mange-tout types. Hold the stem in one hand and snap the pod off with the other, otherwise the whole plant is pulled out of the ground. Pick frequently to ensure continuity of supply

Troubles pea-moth, the white maggot of which eats into peas in the pod, worst on late sowings, spray derris at flowering; greenfly and thrips in dry weather; thrips produce silvery patches on pods and yield is reduced; mildew; slugs; birds; mice, sow seed in containers

Mange-tout

and crop more heavily, but are less hardy and should only be sown from mid spring onwards. There are also the 'petit pois', which naturally have small, sweet peas and are dwarf-growing, and the mange-tout – snow-peas, eat-all, Chinese peas, sugar-peas – easily grown and, again, consisting of different varieties.

Nutritional value

Peas, as well as being a delight to eat, are high in the food value charts, so high in fact that much research is currently in progress to discover ways of including them in cereal products to provide extra protein, or as thickening for gravies and soups to replace flour, since some 50% of dried peas is starch. They contain much more protein and dietary fibre than most vegetables, good quantities of vitamins A and C and a variety of mineral salts, particularly potassium and phosphorus. They are essential in the vegetarian's diet, because of their protein content.

Culinary uses

Dried peas have long been part of man's diet but small tender garden peas were only developed in the 16th century. Simply boiled and tossed in butter they are delicious. Additional flavour can be added by combining them with bacon or onions. Pea soups also have a long history. Mange-tout are best when lightly steamed until just tender but still crisp or stir-fry as they are in Oriental cuisines.

Cultivation

In growing peas in the garden, allowance should be made for only a moderate rate of germination by sowing extra seeds at the end of the rows. Alternatively, peas can be sown in containers and the seedlings transplanted to form complete rows from the start. This also overcomes the problem of cold and/or soggy soil at the time of sowing. Protection from birds, and firm supports are essential, and the soil should be deep, moist and fertile to produce the heaviest crops. For heavy crops, look for varieties on which the pods are in pairs or threes.

RECIPE
Economical

Petits pois, French style
For a more substantial dish, add 250g/½lb streaky bacon (rinds removed) with the onions and serve with triangles of fried bread.
750g/1½lb shelled petits pois
50g/2oz butter
250g/½lb pickling onions, peeled
2tbls chopped parsley
1 small lettuce, shredded
1 or 2 lumps sugar
salt and freshly ground black
*   pepper*
Melt the butter in a heavy saucepan, add the onions and cook gently for about 15 to 20 minutes until onions are soft and lightly coloured. Wash the lettuce and peas and add to the saucepan. Cook covered, on a very low heat for about 20 minutes, adding a little water if necessary to stop the vegetables burning. Add sugar after about 10 minutes and stir vegetables occasionally so they will cook evenly. At the end of cooking time the vegetables should be whole but soft and almost no liquid left. Season with salt and pepper, transfer to a serving dish and stir in chopped parsley.

ESSENTIAL FACTS

Type of plant annual
Part eaten seed, pod
In season early summer–early autumn
Yield 250g (½lb)/plant
Time from sowing to harvesting 11–14 weeks for autumn sown
Hardiness round-seeded hardy; remainder half-hardy
Seed viable 2 years
Germination period 7–20 days

Pepper, Sweet

(Bell pepper capsicum, *Capsicum annuum*) SOLANACEAE

Fundamentally tropical plants, sweet peppers are natives of tropical America, and were introduced to Europe from Mexico by the Spaniards early in the 1500s. They took several centuries to spread throughout Europe, but by the middle of the 19th century were grown regularly in France, Spain and Italy. They have always been widely grown in the warmer parts of America, but it was not until the post war years that they began to be popular in Britain, where they are now an established part of everyday diet.

The sweet peppers are so called to differentiate them from the other large group of peppers derived from *C. frutescens*, which have an exceedingly hot, peppery taste. Sweet peppers have a mildly spicy flavour with overtones of sugar, and a freshly picked specimen is surprisingly juicy as well. They grow into comparatively large fruit, varying in shape from long and conical to square and almost round; the long ones can grow to as much as 25cm (10in), but those in general cultivation are much smaller. The plants grow quickly and carry most of their leaves near the top of the plant, and white flowers appear in midsummer in cool temperate climates, earlier in warm to tropical conditions. Fruit sets easily and prolifically, and often has to be thinned.

Nutritional value

Sweet peppers contain a good deal of nutritive value, as they have a full complement of minerals, some protein and dietary fibre, and an astonishingly high quantity of vitamin C; carotene is not much less and can be a good deal higher, and there are also small amounts of the vitamin B complex. Their water-holding capacity makes them a good summer vegetable, while their vitamin C content ensures their usefulness for the winter diet, especially when used raw in salads.

Culinary uses

They are equally delicious whether cooked or not. The flavour and texture of the cooked pepper is a little different to that of the raw one, which adds a distinctive crispness to a recipe. They can be sliced and used with casseroles, with dishes which include minced meat, chopped and mixed with rice, fried as a side vegetable, or cut into chunks for kebabs, along with meat, tomatoes and onions. Stuffed with rice, tomatoes and minced lamb or chicken, then baked in the oven, they make a filling and delicious main course. Raw, they can be cut into rings for salads or as accompaniments to curries, used in starters and as garnishes. Whatever the use, it is necessary to remove the seeds and inside membrane first.

Cultivation

In spite of their tropical origin, cultivation is not difficult in cool temperate climates. In warm temperate regions, they can be sown and grown outdoors, but in cooler areas they are better grown under cover, and treated more or less the same way as tomatoes, grown in containers, or in a greenhouse border. In hot weather they need a lot of water, and throughout their lives, there should be plenty of plant food available, either in the compost or supplied as extra feeding. In good light and high temperatures they grow rapidly, and crop heavily.

CULTIVATION

Site and soil sunny/as good light as possible, sheltered; well-drained, medium, fertile, manured for previous crop; proprietary potting compost

Soil preparation outdoors, dig 1 spade deep in spring, and mix in fertilizer 2 weeks before planting

Sow indoors, in 18°C (65°F), in late winter–mid-spring, depending on heat supplies available and mildness of garden; sow in seed-trays, spacing seeds 2.5cm (1in) apart, or sow 2–3 in 5cm (2in) pots; sow 0.6cm (¼in) deep

Prick out/thin prick out when 3 leaves present into 5cm (2in) pots; thin to best seedlings per pot; keep temperature at minimum 16°C (60°F) at night and 19°C (68°F) day until in final pots

Pot pot as plants grow, into 7cm (3in), 12cm (5in) and finally into 23cm (9in) pots or grow-bags if to be cropped in a container

Plant outdoors early summer, harden off first, space 45cm (1½ft) apart each way, water in; indoors without heat, late spring, same spacing

Care to encourage side growths, and continued fruiting, take out first flower at top of stem, but if forming bushy growth, leave it; keep moist at all times, especially when fruiting, mulch plants grown in soil; liquid-feed container-grown plants when fruit starts swelling

Pruning/training plants may need a cane or stake to support when in full fruit, as become top heavy

Harvest cut fruits when they are about 7cm (3in) diameter; can be cut at any stage of colouring from green to red, but if left on plant to turn red, other fruits will not be produced and season will be short; if red ones are wanted, they take 2–3 weeks to colour from time swelling ceases, or can be cut and left in warm place to colour

Troubles greenfly; whitefly; red spider mite; caterpillars; sun scald (brown patches) on fruit; grey mould

ESSENTIAL FACTS

Type of plant perennial, grown as annual
Part eaten fruit
In season cool area, midsummer–mid autumn; warm areas, early summer onwards
Yield 4–8 fruits/plant
Time from sowing to harvest 16–20 weeks
Size fruit, 7.5–15cm (3–6in) long, 5–7.5cm (2–3in) wide; plants 60–90cm (2–3ft) tall
Hardiness tender
Seed viable 4 years
Germination period 14–20 days

RECIPE
Healthy

Grilled peppers en salade
2 each red, green and yellow peppers
4tbls olive oil
2tbls wine vinegar
1 tsp Dijon mustard
pinch sugar
salt and freshly ground pepper
1 small can anchovy fillets, drained
stoned black olives, to garnish
quartered hard-boiled eggs, to garnish (optional)

Put peppers under a hot grill or in a hot oven turning frequently until the skins blacken and blister. Then rub off the skins under running water. Halve the peppers, remove the cores and seeds and slice thinly. Arrange on a serving plate. Make a vinaigrette with oil, vinegar, mustard and seasonings and pour over peppers. On top, arrange a lattice of anchovy fillets (if necessary, split down lengthways) and put a stoned olive in the centre of each gap. Arrange quartered hard boiled eggs around the edge of the plate and scatter with fresh chopped parsley or basil, if desired.

Potato

(Batata, Solanum tuberosum) SOLANACEAE

The potato has been part of the staple food diet of Europeans, North Americans and Australians for nearly 200 years; for South Americans the time span is more like 2,000 years. The discovery of South America inevitably meant that the plant with the curious swollen roots eaten by the natives would find its way to the Old World. Its route to Europe is said to have been either by a monk returning to Spain from Peru after the Spanish had overrun the Inca empire in 1531, or by a botanist called Hariot who came back to Britain in a Drake ship, and passed the tubers to Sir Walter Raleigh, when they were planted on his estate in southern Ireland.

They were regarded more as curiosities than an essential food for many years; they had deep eyes and varied enormously in shape and nobbles. But the Napoleonic wars necessitated their growing in large quantities by British farmers, to such an extent that when blight ruined the Irish crop in 1845-46, Ireland was in a state of famine, and mass emigration to America ensued.

The potato plant is a leafy one, with white or purple flowers like those of tomatoes, followed by shiny round green berries late in summer. The tubers are root tubers, with white flesh, provided they are covered with soil, but green when exposed to light; this green part is poisonous and should not be eaten.

Varieties

Potatoes can be divided into two main groups: earlies, and maincrops, with a small section of second earlies, cropping between the two. There are many, many varieties, though commercial growers stick to only about six or seven, but the gardener has different considerations, flavour playing a much larger part, and he or she can experiment with quite different kinds suited to home gardening and family needs. Maincrops take up quite a lot of room if they are to be grown to feed a family until the new season's harvest, earlies are much less space-consuming, and will provide tubers for two-three months.

Nutritional value

Potatoes contain protein, appreciable amounts of vitamin C, and dietary fibre. The calorie count is four times less than bread or fried onions, and is lower than some varieties of bean.

CULTIVATION

Site and soil sun/little shade, not a frost trap; moist soils, manured for previous crop

Soil preparation dig deeply in winter, do not add lime even if acid; fork in a potash-high fertilizer 2 weeks or so before planting

Sprouting use small potatoes the size of a hen's egg – these are 'seed potatoes', and stand them heel (stalk) end downwards in a light frost free place in mid–late winter for earlies, late winter for maincrops; use eggboxes or seed trays. The 'eyes' will sprout shoots

Planting plant each set (seed potato) when shoots are 2.5cm (1in) long, in early spring for earlies, mid-spring for maincrops, end of early spring for second-earlies. Plant in V-shaped drills 10–15cm (4–6in) deep if heavy soil, 15cm (6in) if light; space 30 × 60cm (1 × 2ft) apart for earlies, 38 × 75cm (15 × 30in) for maincrops. Cover drill with soil

Care protect shoots with straw, or bracken, netting or sacking as they appear through soil, if frost is likely; keep weeds under control; earth up stems when 23cm (9in) tall by drawing up crumbly soil from sides to cover stems 12cm (5in) deep, leaving tops protruding and make flat topped ridge. Repeat earthing-up 3 weeks later. Water well in dry weather – do not ever allow plants to run short of water. Remove flowers if they appear. Spray to protect against blight on maincrops early in midsummer and at 2-week intervals until the end of late summer, in damp seasons and where blight occurred the previous year

Harvest dig tubers when haulm begins to yellow; dig earlies as required day by day from early–late summer; dig maincrops all at once late summer–early autumn, leave to dry outdoors or under cover if wet; clean, discard damaged or diseased tubers, and store remainder in dry, dark, well-ventilated but frost-free shed. Stack in layers in boxes. Prevent entry of mice or rats

Troubles potato blight, fungus disease, brown blotches on leaves, stems and tubers, haulm killed, crop reduced; remove and destroy worst affected and spray protective fungicide such as mancozeb every 2 weeks on remainder until lifting. Common scab, brown corky patches on skin, put grass cuttings along drill in next planting season. Underground slugs, eat holes in tubers in late summer and autumn; dig maincrop tubers as soon as possible, apply liquid slug killer midsummer. Cracking, hollow heart, irregular water supplies, keep well watered in dry weather. Wireworm, shiny, yellow, worm-like pests feed inside tubers leaving narrow tunnels, troublesome in ground which was previously turf; no cure, but dig ground over after lifting and expose to birds and weather

Majestic

Pentland crown

ESSENTIAL FACTS

Type of plant perennial, grown as annual
Part eaten root tuber
In season fresh: early summer–mid-autumn; stored: late autumn–mid-spring
Yield 750g (1½lb)/plant, early variety; 1250g (2½lb)/plant, maincrop
Time from planting to harvest 12 weeks earlies, 16 weeks maincrop
Size plants 60cm × 45cm (2 × 1½ft); tubers 25mm (1in)–15cm (6in)
Hardiness tender
Weight of tubers for planting 750g (1½lb)/3m (10ft) row

Desiree

Pentland javelin

Culinary uses

The potato is such an everyday article of food that, from the culinary viewpoint, it often gets treated without interest, possibly because of its association in the past with cheap food for the poor. It has, however, provided the basis for many of the world's classic dishes and can be treated in a number of delicious and by no means plebeian ways.

But it is well worth making it the main item in a meal, and there are many ways of doing this; boiling, mashing, frying and baking are simply ways of serving it as an adjunct, but baked potatoes with stuffings of various kinds make a delicious and nutritious main course, as do gratins, where potatoes can be layered with a variety of ingredients, topped with breadcrumbs or cheese and baked in the oven; there is a tremendous variety of main course salads based on potatoes, many soups including potato soup itself, omelettes and pies, cakes and scones – the list is endless.

Cultivation

Cultivation is straightforward once past the initial stages, when the 'seed' potato must be artificially started into growth, then planted and protected from frost. Earthing up the rows encourages a heavier crop, prevents the tubers from becoming green and supports the haulm. Potato blight, a fungus disease, has to be watched for as it can kill plants, and can infect from nearby infected tomato plants; in summers when prolonged dry weather is followed by heavy rain damaged tubers will result.

Estima

King Edward

RECIPE
Economical

Potato and anchovy gratin
*1kg/2lb potatoes, peeled and cut
    into slices
75g/2½oz butter
2tbls vegetable oil
3 large onions, thinly sliced
15–16 anchovy fillets
salt and pepper
fine white breadcrumbs
150ml/¼ pint single cream
150ml/¼ pint milk*
Heat the oil and 25g/1oz of butter,
add the onions and gently cook until
soft but not brown. Grease a gratin
or baking dish. Arrange a layer of
potatoes on the bottom of the dish,
then alternate layers of onions,
anchovies in a criss-cross pattern
and potatoes finishing with a layer of
potatoes. Season each layer with a
little pepper and a hint of salt.
Scatter breadcrumbs over the top
layer and dot with remaining butter.
Slowly bring the cream and milk to
the boil in a saucepan and pour it
gently down the sides of the dish.
Bake in a preheated oven 200°C/
400°F/gas mark 6, 45 minutes–1 hour
or until potatoes are tender.

Red King Edward

Wilja

Pumpkin/Squash

(Cucurbita pepo) CUCURBITACEAE

Pumpkins are almost certainly natives of tropical America and, if allowed, will grow into trailing plants 2.6m (11ft) long. They were grown and eaten by the South American Indians, and their cultivation in that continent probably extends back much further, for hundreds of years. By the 17th century they were being much used for pies in Britain, made by stuffing the centres with apples; in 1824, reference was made in a book of the time to a pumpkin weighing 120kg (245lb). Cushaw pumpkins, grown in America, are *C. mixta*, but need a much longer growing season, so are not suitable for Britain.

The pumpkin, squash and the marrow (summer squash) are varieties of the same species of the cucurbit family, to which cucumbers, melons and ornamental gourds also belong. But whereas the marrow is generally long rather than round, and striped green with a soft rind, the typical pumpkin is round, with a grooved hard, yellow or orange rind and grows to a vast size. Fruit weighing more than 50kg (112lb) are well within the home gardener's skill in cultivation. Fruit weighing over 200kg (500lb) are possible from an American variety, but they do need a tropical, or at least warm temperate climate, to produce these monsters, and in cooler climates gardeners will have to settle for the comparatively lightweight examples quoted below.

But if several fruits are allowed to set and grow on one plant, 7kg (15lb) per fruit is a more likely weight, and one plant is therefore more than ample for a family. The fruit are left on the plant as long as possible before the first frost, and will then store in good condition for 2–3 months, provided the rind is hard at the time of cutting.

Nutritional value

Pumpkins contain a full complement of minerals, some dietary fibre and protein, and have a low calorie content. Water occupies most of their bulk; the vitamin content is outstanding for the quantity of carotene present, and there is also some vitamin C.

Culinary uses

Pumkins are fun to grow and are always popular at Hallowe'en time; the vast amount of flesh produced can easily be used up as there are many ways to cook it, both sweet and savoury, but for the best flavour, cut pumpkins when only about 23–30cm (9–12in) diameter; as with any vegetable, the smaller they are, the more toothsome, and the rind will be easier to penetrate, too. Use the flesh in jam, chutney, pies and tarts, or in soup and casseroles, as fritters, and as a side vegetable, preferably steamed for the best flavour. The seeds can be eaten after frying in deep oil and then salting, when they are called *pepitos* and served as nibbles with aperitifs.

Cultivation

Cultivation is more or less the same as that for marrow, except that the fruit may have to be left on the plant until mid autumn, in order to mature completely. There is a belief in some areas that pumpkins will grow to enormous proportions if planted amongst marrows; the association somehow seems to be beneficial and results in fruit about three times the size that would normally be expected by home growing.

CULTIVATION

Site and soil sunny sheltered; deep, rich, moist, well-drained

Soil preparation in spring dig a hole for each plant about 38cm (15in) deep and 120cm (4ft) square, and fill with very well rotted organic matter, then cover with topsoil

Sow in mid–late spring indoors, temperature 21°C (70°F); sow singly in 9cm (3½in) pots; sow outdoors on prepared site in late spring and protect with cloche/frame; sow 2.5cm (1in) deep

Pot when roots fill pot, transfer to 12cm (5in) pot, keep at minimum of 16°C (60°F)

Plant late spring–early summer and protect until frost risk is past; space plants 90–120cm (3–4ft) apart

Care water the plants in prolonged dry weather; stop the main stem at about 180cm (6ft) and stop sideshoots to keep within space available; hand-pollinate if fruit not setting by early in late summer; allow about 4 fruits per plant, more if small fruit wanted, less for exhibition specimens

Harvest cut fruit for best flavour when about 23–30cm (9–12in) diameter while rind still soft; cut last fruit before frost and store

Storage hang in a cool, dry, airy place in a strong net or bag and keep for Thanksgiving or Christmas

Troubles mice, eat seeds sown outdoors, flesh of swelling fruit; caterpillars; greenfly; mildew; grey mould

Table ace

ESSENTIAL FACTS

Type of plant annual
Part eaten fruit
In season early autumn–end of mid-winter
Yield average 4 fruits/plant, each about 7kg (15lb)
Time from sowing to harvest 16–18 weeks
Hardiness tender, but less so than melon
Seed viable 4 years
Germination period 5 days; 10–15 at 16°C (60°F)

RECIPES
Economical

Pumpkin Pie
1 medium size pumpkin
3 eggs
milk
demerara sugar
1/4tspn grated lemon rind
125g/4oz plain flour (wholemeal)
25g/1oz butter/margarine
cold water to mix
1 level tspn cinnamon
1 level tspn ginger
baked pie crust
Remove the rind and seeds, slice the pumpkin and boil until tender; liquidize the flesh, then mix in the beaten eggs, cinnamon, ginger and lemon. Sweeten with demerara sugar to taste and add sufficient milk to produce a thick batter consistency. Put this mixture into the baked pie crust and bake for about 3/4 hour at 180°C/350°F/gas mark 4. Serve hot or cold, with cream.

Spiced pumpkin soup
1kg/2lb pumpkin
50g/2oz butter
1 large onion, finely chopped
1 medium potato, peeled and diced
12 coriander seeds, crushed
large pinch cumin
salt and ground pepper
900ml/1 1/2 pints chicken stock
cream and chopped parsley, to garnish
croûtons (optional)
Melt the butter in a saucepan, cook onion until soft but not browned, add pumpkin flesh, potato, spices, seasonings and boiling stock and simmer for 20–30 minutes until vegetables are tender. Liquidize the soup in a blender. Return to the pan and heat through gently. Serve the soup with cream and chopped parsley or, if desired, crisply fried bacon, crumbled, and croûtons fried in butter.

SQUASH
(Cucurbita pepo, C. maxima)

Squashes are prepared and eaten in the same way as pumpkins or marrows. Cultivation is the same, allowance being made in cool temperate climates for the longer growing season needed, by starting them earlier, and cloching them in autumn to complete ripening.

Varieties
There are two sorts of squashes, the summer and the winter variety, both of which belong to the same family as cucumbers and melons. The summer squash (*C. pepo*) group does not store in winter and matures earlier; it can be completely eaten when young, seeds and skin as well as flesh, and it includes the marrow, sometimes also called vegetable squash. Others in the group are the crookneck, yellow or orange skin and shaped like a hook, and the scalloped custard squash like a flattened sphere with scalloped edges, coloured yellow or cream. Pumpkins are also included in this group, in spite of having a hard skin, as they can be eaten when young and indeed are thought to be better flavoured at that stage. The pear-shaped chayote (choko) which flourishes in tropical and sub-tropical countries is also a member of the squash family and can be cooked in the same way as summer squash.

The winter squashes (*C. maxima*) need longer in which to grow and mature, are softly hairy rather than bristly hairy, and have shallowly, not deeply-lobed leaves. They are not eaten young, and mature in autumn, to store well for three or four months. The Hubbard squash, tapering or oval with a warty orange rind, is a popular variety and the turban squash, (*C. m.* 'Turbaniformis'), shaped as its name suggests, is much smaller and coloured red and yellow. The acorn or des Moines squash has a dark green, ribbed rind and grows about 15cm (6in) long and about half as wide; the yellow butternut is a winter squash which is becoming very popular.

Nutritional value
The winter squashes are more nutritious than the summer ones, as they contain more vitamin C, protein, fat and carbohydrate and are less watery

Butternut

Acorn squash

Hubbard squash

RECIPE
Exotic

Oriental squash and pepper salad

*500g/1lb butternut squash or
   pumpkin, cut into slices*
*1 green and 1 red pepper, cut
   lengthwise into strips*
1½tsps cumin seed
5tbls fresh orange juice
2tbls fresh lime or lemon juice
2 spring onions, finely chopped
1tbls fresh mint, finely chopped
3tsps sugar
*3tbls toasted cashew nuts or
   almonds, coarsely chopped*

Steam the squash or pumpkin until
just tender but still firm. Allow to
cool and arrange on a dish with the
peppers. Toast the cumin seed in a
dry saucepan, taking care it does
not burn, then crush with a mortar or
rolling pin. Combine the orange
juice, lime juice, spring onions, mint
and sugar. Add the crushed cumin
and blend well. Pour dressing over
vegetables, toss gently and sprinkle
with toasted nuts.

CULTIVATION

Site and soil sunny, or slightly shaded for high summer sowings; light, sandy, fertile soil ideal, but will grow on most

Soil preparation dig one spade deep and mix really well rotted organic matter in during winter; fork in fertilizer 2 weeks before sowing; prepare seed-bed to very fine tilth otherwise germination poor

Sow summer varieties outdoors from early spring to late summer, in succession, sow thinly 2.5cm (1in) apart in rows 15cm (6in) apart; winter varieties outdoors mid–late summer, also thinly, in rows 30cm (1ft) apart; sow 1.5cm (½in) deep

Thin summer varieties to 7cm (3in) apart, do as soon as possible after germination; thin winter varieties to 15–23cm (6–9in) apart

Cultivation keep weeds under control; always water well in dry weather to keep plants growing without check, otherwise woody and flavourless

Harvest pull roots when round summer ones about 2.5cm (1in) in diameter, or if long, about 5cm (2in); leave winter kinds where they are until wanted, protect from frost with bracken or straw

Troubles flea-beetle; birds attacking seedlings; drought; occasionally troubles as for other brassicas

Radish

(Raphanus sativus) CRUCIFERAE

The word 'radish' is derived from the Latin word *radix* meaning root (cf radicle); *raphanos* means 'easily grown', and these two meanings sum up the essence of the radish. It is, indeed, an easily grown root, whose beginnings are lost in the mists of antiquity. The ancient Egyptians knew it, and there is a theory that it was first cultivated in China and Japan – another is that it was derived from the wild radish but that seems a little far-fetched, as the wild species is endemic to Europe.

It has always been a popular salad vegetable and there are mentions of it in records from the Roman invasion of Britain to the present day. It was once thought to be an antidote to mushroom 'poisoning' and to have many and varied medical uses. In mediaeval times it was recommended for eating before meat as what would now be called a starter, or 'with meates to procure appetite', presumably with the same intention.

Varieties
The small round or tapered varieties with white flesh and red or white skins are familiar enough, but there are also large, white-skinned, long radish varieties for summer, and large white or red radishes for winter cropping, together with 'black' or brown radishes, round, cylindrical or long. Their size is much greater, as the average weight can be ½–1kg (1–2lb) and there are some Japanese kinds which can weigh an incredible 18kg (40lb), but nothing is new under the sun – monsters of this kind were described in the literature in 1552.

Nutritional value
Nutritionally they have a low calorie content, but a surprising amount of iron, a variety of other minerals and a lot of vitamin C, so are a good vegetable for winter salads. The high water content makes them a useful thirst-quencher in summer.

Culinary uses
The combination of crispness and succulence, together with their hot flavour, must make radishes unique amongst the vegetables, and it is these qualities which add so much to a green salad. However, while they are generally eaten raw, they can be boiled and served as a hot, cooked vegetable, and this particularly applies to the winter varieties. The seedpods can also be eaten, and some varieties are grown specially to provide pods, eaten raw or pickled in salt water and vinegar.

Cultivation
Radishes grow quickly and can be ready for pulling within a few weeks during spring and autumn; the winter kind take longer to mature but will then stand in the ground for several months without deteriorating. Summer radishes have a tendency to run to seed and need to be sown in shade and kept well watered, but even if they do, the seedpods are edible. Flea-beetles can be a problem, feeding on the leaves of seedlings such that they can kill them if not controlled.

Long black Spanish

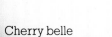
Cherry belle

Scarlet globe

Winter radish

ESSENTIAL FACTS

Type of plant annual or biennial
Part eaten root
In season mid spring to late winter
Yield summer varieties: 2kg (4½lb)/
3m (10ft) row; winter varieties: 5kg
(10lb)/3m (10ft) row
Time from sowing to harvest
summer, 3–6 weeks; winter, 12–14
weeks
Size roots: summer, 2.5 × 2.5–5cm (1
× 1–2in); winter, 15–30 × 5–7.5cm
(6–12 × 2–3in)
Hardiness hardy
Seed viable 4 years
Germination 4–10 days

RECIPE
Healthy

Radish and orange salad
1 winter radish, thinly sliced
1 orange, thinly sliced
lemon juice
coarsely ground salt and pepper
Arrange the orange and radish
slices alternating on a serving
platter. Sprinkle with lemon juice,
salt and pepper. Serve with slices of
dark rye bread, if desired.

Rhubarb

(Rheum rhaponticum) POLYGONACEAE

There are both medicinal and garden forms of rhubarb, the medicinal one, *Rheum officinale*, being recorded in Chinese medicine as long ago as 2,700 B.C. It was eventually imported by the Greeks and used by the Greek doctor Dioscorides, and subsequently spread throughout Europe, albeit rather slowly. The garden rhubarb was first cultivated in Padua early in the 17th century, for its tasty young stems, and from there it spread to Britain where it became very popular in Victorian times, so much so that there were over 100 different varieties – a far cry from today, when there are only half a dozen or so available.

Nutritional value

The large leaves and thick red or green stems are familiar enough to need no further description, but it should be remembered that the leaves are harmful and not to be eaten. The stems contain oxalic acid, like spinach, which renders the calcium unavailable, but adding milk to pastry or steamed puddings, or serving it with rice pudding, custard or cream will offset this lack. The calorie count is low, there is a high water content, and some dietary fibre, together with vitamins A and C, potassium and other mineral salts. Fat and protein are almost non-existent.

Culinary uses

Although essentially a vegetable, the stems of rhubarb are treated as a fruit, and they can be stewed or baked in the oven, made into pies, puddings or crumbles and served as a dessert. They can be preserved by bottling, freezing or making into jams or chutneys. Because of their high water content, the young stems lend themselves to purées to be mixed with cream or custard. Rhubarb should not be used after the end of spring, as the stems become tough and tasteless, though research being carried on at present has produced varieties whose stems can still be pulled throughout early summer.

Cultivation

As regards cultivation, rhubarb can be left pretty much to its own devices, once planted, provided it has a good strong soil and is generously mulched every year. Although it can be grown from seed, it is a slow and unreliable method, and plants bought from a garden centre or nursery provide an easier method of starting. Plants can live 20 years or more, but if the supply of stems begins to diminish, dig up and divide a crown or two and replant the best divisions with buds attached. You may need a saw to divide the plant.

CULTIVATION

Site and soil sun or shade; deep moist, fertile soil

Soil preparation dig deeply autumn/winter, mix in organic matter at the same time; fork in bonemeal and woodash a few days before planting

Plant late autumn or early spring, spread roots out fully and cover crown with 5cm (2in) soil. Space 90cm (36in) apart each way

Care keep free of weeds while young; mulch midsummer when harvesting finished; water well in very dry weather

Prune/train cut off flowering stems when still in bud

Harvest do not pull stems 1st summer after planting; then pull a few from each plant at a time; pull so that whole stem plus basal bud comes away, trim off leaves and put on compost heap

Forcing rhubarb can be forced by covering the plants in early winter with tubs or boxes at least 60cm (2ft) deep, when it will be about 5 weeks earlier, and then cover with garden compost. Do not force again for at least 2 years

Troubles crown rot, an infrequent fungal infection, stems poor colour, root rots, no cure, destroy and plant in different site

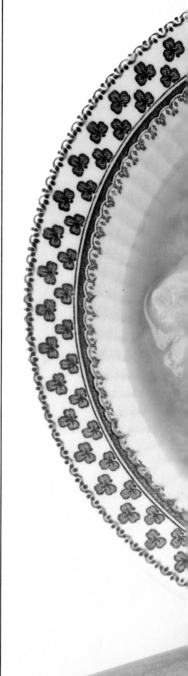

Type of plant herbaceous
perennial
Part eaten leaf stem
In season mid-spring–early
summer
Yield average 3kg (6lb) per plant/
season
Time from planting to harvest 12–
18 months
Size 60 × 90cm (2 × 3ft) without
flower stems
Hardiness completely hardy
Seed viable 3 years
Germination 2–3 weeks

RECIPE
Economical

Rhubarb sauce (to serve with
grilled mackerel)
500g/1lb fresh rhubarb
125g/4oz demerara sugar
2tbspn water
¼tsp lemon juice
Skin and chop rhubarb into 2.5cm
(1in) lengths, place in ovenproof
dish with water and demerara
sugar, cover, and cook in slow oven
until cooked but pieces still whole,
about 30 minutes. Cool and mash or
purée with the lemon juice. The
sauce should be slightly tart and not
too sweet.

Spinach

(Spinacea oleracea (winter spinach), *S. o.* 'Inermis' (summer spinach); *Beta vulgaris* 'Cicla' (spinach beet/perpetual spinach); *Tetragonia expansa* (New Zealand spinach))*
CHENOPODIACEAE, AIZOACEAE

Spinach of some kind was listed as a garden plant as long ago as the 15th century in Britain, when it was used in sweet dishes. In spite of this, it was known as the 'Spanish vegetable' because it was thought to originate in Spain, and was introduced by the Moors. In fact, Asia is its native habitat, and it was eaten by the Greeks and Romans, not just used for medical purposes. In 1578 an English writer described it as a 'salat-herbe'; a hundred years later it was used for stomach troubles. Parkinson, the 17th century gardening writer commented that the Dutch cooked it without water, but its widespread use did not occur until the 19th century, in the Netherlands, France and England, later the rest of Europe and the Americas.

Varieties

There are several sorts of spinach, all grown from seed, sown any time from spring until mid-late summer, but mid-spring and midsummer are the most popular times. All have the same flavour and, partly because of its lasting qualities and partly because it is unlikely to bolt, perpetual spinach is the most widely grown. It is a type of beetroot, but with a branched root system, and large leaves with prominent midribs growing in a cluster at ground level.

Winter and summer spinach have smaller, paler, thinner leaves and round seeds in the case of summer, prickly, 3-pointed seeds in that of winter spinach. Summer spinach is very inclined to bolt in summer, long before any quantity of leaf is picked. New Zealand spinach is quite different to the others as it has long stems which trail along the ground and small, thick grey-green leaves all along the stems. The shoot tips can be eaten as well, and their removal encourages sideshoots to grow. It tolerates dry conditions well, but is tender and will be killed by frost.

Nutritional value

From the health point of view spinach contains a good deal of vitamin A and B, a useful quantity of vitamin C, together with potassium, iron and calcium, but most of these are lost if the leaves are boiled. They are better steamed, but even so the water used for steaming becomes discoloured, suggesting that some goodness is still being lost. The oxalic acid in spinach makes its calcium unavailable when boiled in water. To retain its nutrients, it is best simmered in milk as a soup. With a low calorie value and high dietary fibre, spinach has a great deal to commend it as a vegetable: easily grown, and prolific in crop, it has considerable dietetic use.

Culinary uses

Spinach provides a base for a great variety of dishes, such as soufflés, soup, salads, lasagne verde, omelettes, tarts, purées, and egg and fish dishes; many with the word Florentine in the title have spinach in them. The leaves should be thoroughly washed before cooking. Ideally, it should be gently cooked without water, moisture being provided by water clinging to the leaves after washing.

Cultivation

Spinach is particularly easily grown, and has hardly any problems. A moist, slightly heavy soil will ensure a prolific yield, giving it a high value for the space it takes up. In dry conditions or light soils, it must be watered well and

CULTIVATION

Site and soil open/slightly shaded, moist slightly heavy-medium soil, neutral-alkaline

Soil preparation dig in manure/compost in autumn/early winter; fork in general compound fertiliser 10 days before sowing; add lime in winter if soil acid, at an interval of at least six weeks from manuring

Sow summer spinach and perpetual spacing seeds thinly and evenly in rows 30–38cm (12–15in) apart; perpetual spinach and winter spinach late summer at 15cm (6in) stations; New Zealand spinach late spring, 2cm (¾in) deep, sow few seeds 30cm (12in) apart

Thin spinach 23–30cm (9–12in) apart, perpetual spinach 30cm (12in) apart New Zealand to 60cm (2ft)

Care weed well in early stages and watch for slugs; water during dry weather and watch for greenfly; mulch to keep moisture in soil; protect in winter with cloches against cold and birds

Harvest pull leaves as required; summer spinach late spring–late autumn, winter spinach late autumn–mid-spring; perpetual all year; New Zealand early summer to frost, use shoot tips as well

regularly, especially summer spinach which will otherwise bolt, and if this type unavoidably has to be grown on light soil, it will crop better in light shade. Slugs and snails can decimate the young plants, and birds will feed on the overwintering kinds. Protecting them from cold with cloches will defeat these predators as well.

Winter and summer spinach will between them provide leaves all year round; perpetual spinach will also do this from one spring sowing, but a second one made in mid to late summer will provide better leaves in winter and will overlap a new spring sowing. New Zealand spinach is sown later, in late spring, and will only crop to the first frost. The seeds of all but the last produce several seedlings from each, as the 'seed' is actually a capsule.

ESSENTIAL FACTS

Type of plant annual except perpetual spinach, which is biennial
Part eaten leaf
In season all year
Yield 140g (5oz) per plant summer spinach; 224g (8oz) winter, perpetual; New Zealand 336g (12oz)
Time from sowing to harvest 8–12 weeks
Size 20–30cm (8–12in) high; New Zealand stems 60cm (24in) long
Hardiness all hardy except New Zealand
Seed viable 2 years, except perpetual, 4 years
Germination period 10–20 days

RECIPE
Exotic

Spinach Terrine Parisienne
750g/1½lb spinach leaves, lightly cooked and squeezed of moisture
1 tbls each chopped parsley, chives and tarragon
150ml/¼ pint double or soured cream
3 eggs
75g/3oz each leeks, carrots, red peppers, courgettes and artichoke hearts
Put spinach, herbs, cream and eggs in a blender or food processor and blend until smooth. Add seasonings to taste. Cut the vegetables into strips and cook in boiling salted water until just tender. Drain and set aside. Line a rectangular terrine with buttered greaseproof paper. Fill the terrine in layers, starting with the spinach purée, then the vegetable strips and so on, finishing with spinach mixture. Cover with buttered greaseproof paper or foil and cook in a bain-marie in the oven (180°C/350°F/gas mark 4) for about 1 hour or until set. Leave to cool, then chill in the refrigerator. Serve with a tomato coulis.

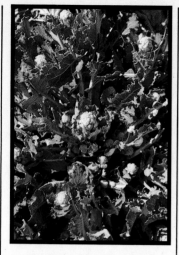

CULTIVATION

Site and soil sunny/open, sheltered; moist soils, well-drained, alkaline, manured for previous crop

Soil preparation dig 2 spades deep in autumn; lime in winter if soil acid; rake in fertilizer 2 weeks before planting, tread soil in advance

Sow outdoors thinly in rows in nursery bed mid–late spring; sow 1.5cm (½in) deep

Thin to 7.5cm (3in) apart, do not allow them to become spindly

Plant early–late summer; water plants before lifting; plant firmly in firm soil so that lower leaves just above soil; plant when 3–4 leaves present; water in; space 60cm (2ft) apart each way

Care watch for slugs/snails and birds on young plants; water in dry weather, and mulch; liquid-feed occasionally; protect in winter from birds

Prune/train earth up stems in autumn; stake in windy areas

Harvest purple sprouting, late winter–mid spring, earlier with early var. and/or mild gardens; white sprouting early–late spring; snap off terminal florets first, then side spears from top downwards, when about 10–12cm (4–5in) long; keep picking, otherwise new spears stop appearing, but do not strip; expect to pick from one plant for about 6 weeks.

Troubles general cabbage troubles but less likely

Sprouting Broccoli

(Brassica oleracea 'Cymosa'*)* CRUCIFERAE

Sprouting broccoli seems to have originated in Italy, being yet another variation on the ubiquitous wild cabbage which has obtained its English name straight from the Italian, as broccoli means stalklets; *brocco* is a shoot. Broccoli from Naples was described by Evelyn the diarist writing at the close of the 17th century; its cultivation was given in detail in an English gardening book widely published in many editions from late in the 18th century, but it was still described in an authoritative book on vegetables published in 1885 as 'the new vegetable' and called sprouting or asparagus broccoli. Now widely grown in Great Britain, it should not be confused with calabrese, which has a different colour, a tightly packed head of florets, crops at a different time of year, and is grown differently.

Sprouting broccoli is an extremely useful vegetable because it provides fresh greenstuff at a time of year seriously short of fresh green vegetables; not only that, but young broccoli spears are deliciously succulent when picked at the right stage of growth, and not left on the plant so long that they become fibrous and coarse-flavoured.

Varieties

This broccoli is called sprouting to differentiate it from winter broccoli, which looks so like cauliflower and to all intents and purposes is one, that it is generally called a cauliflower. Sprouting broccoli is often referred to simply as 'sprouting'. It is grown for its small heads of florets which grow on sideshoots all the way down the main stem, appearing in the second year of its growth from early spring onwards. The florets can be purple or white, depending on variety, the white being less hardy but better-flavoured. There is a variety called 'Nine-star Perennial' which has a central head with smaller clusters around it in early–mid-spring; providing all the florets are taken, the plant will continue for several years, gradually getting larger. It needs a fair amount of space each way and should be well mulched each spring after cropping.

Nutritional value

Sprouting broccoli is also highly nutritious, containing a good deal of protein, high counts of minerals and vitamins, some dietary fibre, and a low calorie value.

Culinary uses

Steaming is preferrable to the standard method of cooking by boiling, and it can be served as a side vegetable plain or with a white sauce or with butter, included in vegetable soups, curries or vegetable aspics, as a substitute for cauliflower sprigs in recipes for that vegetable, and in salads, either raw and marinated in vinaigrette, or partly cooked and served cold.

Cultivation

Cultivation is similar to the rest of the cabbage family and good crops can be obtained without difficulty, provided the soil is firm, well-drained and reasonably fertile, and the plants have a long growing season. See pages 42–43 for diseases and pests affecting brassicas.

Type of plant biennial
Part eaten young sideshoots and florets
In season late winter–late spring
Yield ½kg (1lb)/plant
Time from sowing to harvest 9–11 months
Size plants 90 × 45cm (3 × 1½ft)
Hardiness hardy
Seed viable 4 years
Germination period 7–12 days

RECIPE
Healthy

Cream of broccoli soup
250g/½lb broccoli spears
1 large onion
1 potato
salt and pepper
25g/1oz butter
450ml/¾ pint chicken stock
150ml/¼ pint single cream
1tbs parsley
nutmeg
sour cream (optional)
Slice the onion and potato and sauté in the butter until the onion is transparent. Add the broccoli, stock and seasonings and simmer for 15–20 minutes until the vegetables are tender. Remove from the heat, add the cream, nutmeg and parsley and liquidize. Reheat gently and serve with sour cream, if desired.

Swede

(Rutabaga, *Brassica napus* 'Napobrassica') CRUCIFERAE

Though some people say that the swede is simply a large turnip, the flavour is quite dissimilar. Swedes provide a winter root crop that is available for many months and survives extreme cold. They were actually introduced to Britain from Sweden as the Swedish turnip (rotabagge), and are thought to have been the result of a cross between *Brassica oleracea* (cabbage) and *B. rapa* (turnip), forming that most unusual hybrid, a cross between two species.

It seems to have first appeared in the 17th century, in Bohemia. In Britain it was not known until the 18th century, and was then grown mainly as food for livestock; by the 19th century it was a regular part of everyday human food, used particularly in pasties, and in mutton stews and hotpots. It is part of the traditional meal eaten by the Scots on Burns Night at the end of January, which consists of haggis, tatties and neeps (potatoes and swedes), and in fact once it was found that swedes could withstand the winters of northern Britain, it was widely adopted and cultivated there.

The swede is a large, hard swollen root, shaped roughly like a long globe, the skin of which can be yellow, white or purple and the flesh yellow or white, cooking to orange or yellow. It grows slowly, and will then stand in the ground without deteriorating through the autumn and winter.

Nutritional value

Nutritionally, its most valuable point is its vitamin C content, but it also has many mineral salts, some calories and dietary fibre, and a small amount of protein.

Culinary uses

Mostly used as a side vegetable it can, however, be associated with other foods in much the same way as parsnips, being steamed or roasted with a joint, or mixed into casseroles or vegetable soups.

Cultivation

Cultivation is of the simplest, and its wants are few; water will always be necessary to prevent the roots being small and woody, but flea beetle can be a serious problem on the seedlings.

CULTIVATION

Site and soil sunny/open; moist soils, not waterlogged in winter, not acid, manured for previous crop

Soil preparation dig 1 spade deep in winter, add lime if soil acid; rake in fertilizer 2 weeks before sowing; firm soil by treading

Sow outdoors where it is to grow, early summer, late spring on cold sites; sow thinly 1.5cm (½in) deep, rows 38cm (15in) apart

Thin to 23cm (9in) apart

Care watch for flea-beetle on seedlings/young plants; water well in dry weather; keep weeded and mulched

Harvest dig up as required from mid-autumn

Store leave in ground all winter, or dig up and store in dry peat in rodent-proof container

Troubles virtually none, but occasionally flea beetle, and swede midge, tiny white maggots which eat growing point of young plants, plant ceases to grow, destroy maggots with g-HCH spray

Type of plant biennial
Part eaten 'root'
In season mid-autumn–spring
Yield average weight/root, 1kg
(2.2lb)
Time from sowing to harvest 20–24
weeks
Size 10–17cm (4–7in) diameter,
12–20cm (5–8in) long
Seed viable 2 years
Germination period 6–10 days

RECIPE
Economical

Bashed neeps
*1kg/2lb swedes (or 500g/1lb each
   potatoes and swedes)*
50g/2oz butter
powdered ginger
nutmeg
salt and pepper
Cut the swedes into chunks and boil
in salted water until tender. Drain
thoroughly before mashing and
drain off any liquid after mashing.
Add the butter, salt and pepper and
a little powdered ginger and
nutmeg. Heat through well before
serving with chopped parsley, if
desired.

Sweetcorn

(Indian corn, maize, *Zea mays* 'Saccharata') GRAMINEAE

Corn or maize was introduced to Europe by Christopher Columbus after his discovery of America, where it had been widely used for both man and beast by the American Indians in central and South America, especially Mexico and Peru, and in southern North America. It was sent back to Seville, in Spain, and from there spread widely throughout Europe and the rest of the Old World, but was used mostly for feeding animals.

Although, in Britain, it was recommended as a garden vegetable in 1829, widespread growth did not begin until after the Second World War, and even now it is not a regular inhabitant of the average British or European vegetable garden.

Sweetcorn is the only member of the grass family grown by the vegetable gardener, and is a native of the sub-tropical parts of America. Maize itself was originally grown to be ground into meal, which was then used in various ways. Present-day use includes the South American and Mexican tortillas for which the meal is formed into cakes and cooked; and a porridge, in which the meal is boiled with water to provide the mealies of South and Central Africa. It is also widely used as a foodstuff for animals, Argentina being one of the chief exporting countries, and in America and parts of France most of the maize produced is used for animals, particularly pigs. The leaves and stems are used as silage. Cornflour is made from maize – it has a high starch content – and corn-oil made from the embryonic kernel has many uses, particularly in cooking and the manufacture of margarine. Cornflakes consist of maize which is pre-cooked, flaked and toasted.

Varieties

There are a number of different varieties of maize: *Z. m.* 'Indurata', or flint corn, provides the corn most used for human foods, except sweetcorn; *Z. m.* 'Indentata' or dent corn is for livestock, *Z. m.* 'Rostrata', for popcorn, and *Z. m.* 'Saccharata', sweetcorn and the only one with wrinkled seeds. The ornamental corns belong to the popcorn group, including the one now called Indian corn with red, black, orange, yellow and cream grains, and strawberry corn, shaped and coloured like a large strawberry.

Nutritional value

Sweetcorn has considerable nutritional value and a delicious flavour. It has high mineral and protein contents, a good deal of dietary fibre, and good quantities of vitamins except nicotinic acid, a member of the vitamin B complex. This is a characteristic of maize in general, hence the prevalence of the B deficiency disease, pellagra, amongst communities which rely too much upon maize meal as a food.

CULTIVATION

Site and soil sunny, sheltered; moist soils, well-drained, slightly acid preferred, but will grow in others
Soil preparation dig 2 spades deep, dig in organic matter; fork in fertilizer about 2 weeks before sowing/planting
Sow indoors mid spring, in 16–18°C (60–65°F); sow 2–3 seeds 2cm (¾in) deep per soil block/peat pot, 7cm (3in) diameter
Thin to 1 seedling, the strongest in each pot
Plant outdoors late spring under cloches; plant pot/block complete 30–45cm (1–1½ft) apart each way, in blocks of threes; harden off before planting
Care remove protection when tips of plants touching top; keep well watered at all times; mulch to cover basal buttress roots when they appear; shake the tassels over the plants to disperse the pollen; supply supports in windy gardens
Harvest test the cobs for ripeness by pulling the silks gently when they have turned brown and dry-looking; if they part easily the cob is ripe, if not, wait a little longer; cream-coloured cobs and kernels producing a milky juice also indicate maturity; water shows under-ripeness, mealiness over-maturity; twist cob off stem and cook immediately, as sugar conversion to starch starts at once; haulm makes excellent compost
Troubles lack of cobs, no pollination, cool conditions, dry soil, late planting; fritfly, tiny, white maggot eats growing point of seedling, cobs small and distorted, look for it in early and mid-late summer and hand pick; smut, fungus disease, large grey or white galls on plants which are distorted including cobs, destroy plants as soon as seen and before galls burst to release spores, do not grow sweetcorn in same site for at least 5 years

Culinary uses

In Britain and most of Europe it is mainly eaten in the form of corn-on-the-cob (whole cobs are boiled and served with melted butter and seasonings). In America, the kernels are detached from the cob and treated in much the same way as peas, being tinned alone or with peppers, and served in salads, rice mixtures, mixed vegetables and fritters. They are also commonly used to make muffins and breads and for chowder and soups.

Cultivation

Sweetcorn is a tall-growing plant needing a warm summer to mature its cobs and to develop several on each plant rather than just one or two. Each cob is produced in the axil of a leaf, with a green sheath-like outer covering, from which protrude the long, thread-like silks, the styles of the female flowers. Higher up the plant, at the end of the stem, are the male flowers, consisting of feathery, branching spikes known as the tassel. Plants are wind-pollinated and best results are usually obtained if they are grown in blocks, or groups of three if space is tight. Shelter and plenty of warmth are great assets. Varieties used not to crop until late summer, and it was not possible to grow sweetcorn in northern Britain or Europe, but there are now early ripening varieties which will also stand much cooler summers, so that growing sweetcorn is less of a gamble than it used to be.

ESSENTIAL FACTS

Type of plant annual
Part eaten seedhead
In season midsummer–early autumn
Yield 2 cobs/plant in cool areas; up to 6 in warm ones
Time from sowing to harvest 12–16 weeks
Size plants 120–180cm (4–6ft) tall; cobs 12–17cm (5–7in) long
Hardiness tender
Seed viable 2 years
Germination period 10–15 days

RECIPE
Exotic

123

Succotash (with chicken)
Serve this traditional American dish with grilled or roast chicken or, as suggested here, in a hearty chicken casserole.

4 cobs of corn, boiled and kernels removed or approx. 500g/1lb corn kernels, canned or frozen
50g/2oz butter
2 onions, chopped
1 clove garlic, finely chopped
300g/10oz broad beans, shelled or frozen
1 sprig of thyme
1 sprig of chervil
150ml/1/4 pint double cream
lemon juice to taste
2tbls parsley, finely chopped
2tbls chives, chopped
150ml/1/4 pint chicken or vegetable stock
1.75kg/31/2lb chicken, cut into serving pieces (optional)
salt and freshly ground pepper
water biscuits or other crackers as an accompaniment

If using the chicken, brown the pieces in a frying pan in the butter mixed with a little oil. Remove and set aside. Pour off all but 1tbls of fat, add the onions and garlic to the frying pan and cook over moderate heat for about 4 minutes until soft but not brown. Add the stock, beans, thyme, chervil, salt and pepper. Return chicken pieces to pan and simmer gently for about 10–15 minutes or until chicken is tender (8 minutes if not using chicken). Transfer chicken to a warmed serving dish and keep warm. Add cream to pan and boil for about 2–3 minutes until mixture has reduced and slightly thickened. Return the chicken to the pan and cook for another 2–3 minutes until chicken is heated through. Add lemon juice, check the seasoning and stir in the parsley and chives. Transfer to a warmed serving dish and serve with water biscuits as an accompaniment.

CULTIVATION

Site and soil sun/little shade, most soils, preferably moist, but well-drained

Soil preparation dig 1 spade deep in late autumn–winter, mix in rotted organic matter; fork in general fertiliser 2 weeks before sowing

Sow white variety late spring outdoors; sow red variety 1 month later to avoid bolting; space individual 'seeds' 10cm (4in) apart and rows 38cm (15in) apart; sow 2.5cm (1in) deep

Thin seedlings at each station to 1, and thin to a final spacing of 30cm (1ft) between plants

Care ward off slugs/snails until well established; keep weeds under control; mulch well; water red variety early in dry weather, white can stand a little longer; protect with cloches, straw or bracken in winter; remove flowerheads before flowering

Harvest cut stems of leaves about 4cm (1½in) above soil level; take 1 from each plant when cutting starts, then gradually increase as plants mature; take outer leaves first and always take them before they are full grown

Troubles slugs/snails

Swiss Chard

(Seakale beet, silver beet, leaf beet, *Beta vulgaris* 'Cicla')
CHENOPODIACEAE

Another vegetable with a long history, Swiss chard was regarded by the Greeks as a food plant, as did the Romans who considered it a delicacy. Since then it has never gone out of cultivation, though with the appearance of cultivated celery, it became less popular. Indeed, many perfectly good vegetables seem to have been discarded in favour of celery, presumably because it survives transportation without losing much of its freshness.

Swiss chard is an excellent vegetable not grown in gardens as widely as its flavour merits. It is a popular vegetable in France and Italy but it does not travel well and takes up a lot of space. Yet it is grown as easily as spinach or spinach beet, which it resembles in flavour, produces more crop for the space, and supplies two types of vegetable: the leaf, and the leaf stem which runs on into the central leaf rib.

Varieties

The leaf stem is broad and flattened, at least 15mm (1in) wide, white in the common variety, deep red in the variety called Ruby or Rhubarb Chard. Swiss Chard – the white variety – stands up to hot dry weather better than summer spinach, though the red kind has a tendency to bolt in the first summer, and needs to be kept well supplied with water. There are other varieties with deep green leaves, and flat or curled leaves like Savoy cabbages, but these are rarely available now.

Nutritional value

Food value figures are not available, but it can be assumed that the leaves contain much the same vitamins and minerals as spinach in the same quantities. The leaf complete with midrib will contain rather more dietary fibre, and proportionately less water.

Culinary uses

The stems are stripped of the leaves and cooked like asparagus, then eaten with melted butter. The leaves can be used in soups, rice dishes, and in general treated like spinach, though if eaten as a straight vegetable, the addition of sorrel leaves is needed to complete the spinach flavour.

Cultivation

The cultivation of Swiss chard is a straightforward affair; once past the vulnerable-to-slugs/snails stage, it can be left to its own devices, apart from watering in dry weather, and even this does not have to be rushed, as with spinach.

ESSENTIAL FACTS

Type of plant biennial
Part eaten leaf, leaf stem
In season midsummer–early winter, mid-spring–early summer
Yield 330g (¾lb) per plant
Time from sowing to harvest 12 weeks
Size plants 30 × 45cm (1 × 1½ft)
Hardiness hardy except in moderate cold, −3°C (25°F) or less
Seed viable 4 years
Germination period 6–14 days

RECIPE
Exotic

Swiss chard with hollandaise sauce

*500g/1lb Swiss chard, stems and
   leaves separated*
3tbls white wine vinegar
3tbls water
3 large egg yolks
175g/6oz butter
salt and lemon juice
chopped chives

Wash the chard stems, cut into small lengths, stripping off stringy parts. Put into boiling water and cook until just tender. In the meantime, make the sauce. Boil vinegar and water until reduced to 1tbls. Leave to cool before beating in the egg yolks. Melt the butter. Set the mixture over a pan of barely simmering water and slowly add the butter beating all the time. Remove when the mixture has slightly thickened, add salt, lemon juice and chives and serve immediately over the chard stems.

Tomato

(Love-apple, apple of Peru, Peruvian apple, *Lycopersicon esculentem*) SOLANACEAE

It is said that tomatoes were first introduced to Europe by the Spanish and Portuguese from Peru in the 16th century, but there is also a reference to their use by the Greek physician Galen in A.D. 200, and there is a sub-species of tomato called *L. e. galenii*, from which the cherry tomato is derived. The botanical name was once applied to an Egyptian plant. Whatever their origin, tomatoes spread rapidly through the Mediterranean region; in Britain they were grown by Gerard, who described them in detail, but made no reference to eating them, beyond saying that: 'in Spaine . . . they doe eate the Apples with oile, vinegre and pepper mixed together for sauce to their meat.' They were grown principally for their decorativeness; in the early days, they were thought to have aphrodisiac qualities.

The tomato is so widely grown now that it is hard to imagine a time when its red fruits were unknown at table, but surprisingly it did not become such a common vegetable in the Western diet until late in the 19th century. Part of the problem of course was that it needs warm summers to ripen a satisfactory crop, and in cool temperate climates, ripening does not start until late summer outdoors, and in a bad summer not until early autumn, thus allowing only a few weeks in which to pick. Now grown under glass, both commercially and by the home gardener, tomato cultivation is as much a part of the growing and gardening scene as potatoes and parsnips.

Nutritional value

Tomatoes are in fact as nutritious as any other vegetable; they contain a good deal of various vitamins, a range of minerals, some dietary fibre, a little protein and much water. The flavour is distinctive, and not universally appreciated by any means – certainly the best flavour is not found in tomatoes picked for more than a few hours, or grown in cool dull conditions. Warmth, sun, and plenty of moisture to ensure full development impart quite a different taste, as those who have sampled Spanish and Italian tomatoes just off the vine and eaten in the open air, will confirm.

Marmande

CULTIVATION

INDOOR

Site and soil sunny; deep, well-drained fertile, alkaline, or use potting compost

Soil preparation dig 2 spades deep, mix in rotted organic matter generously early winter; lime if necessary late winter

Sow late winter–mid-spring, in seed-trays or pans, spaced 2.5cm (1in) apart, temp. 18°C (65°F); sow 1.5cm (½in) deep

Prick out when 1st true leaf starts to appear, into trays or pans spaced 5cm (2in) apart, or singly into 5cm (2in) pots; maintain same temperature; discard seedlings with unevenly-sized seed-leaves; discard those with more closely spaced leaves than majority

Pot into 9cm (3½in) pots when roots fill small pot or leaves touching in tray/pan; space pots out as plants grow so that leaves not touching; night temp. 16°C (60°F); day temp. 18–21°C (65–70°F); if to be grown in pots, pot into 12cm (5in) pots and then final size, 23cm (9in) diameter; use soil-based compost J.I. potting No 3

Plant in greenhouse soil when roots fill 9cm (3½in) pots; 1st truss should be showing; with heat, early in mid-spring, without heat, late in mid-spring–late spring; water plants 12 hours before planting; space 45 × 60cm (1½ × 2ft) apart; put supports in place before planting; water in a circle round root ball, not ball itself

Care keep temperature high, but also maintain good ventilation by use of ventilators and doors; water plants regularly, and once fruit is setting, heavily, particularly in containers; liquid-feed in containers with potash-high fertilizer until late summer, then change to nitrogen-high to keep plants going while they mature rest of crop; spray overhead in mornings while flowering to ensure good and regular fertilization

Prune/support supply 1 cane per plant 180–210cm (6–7ft) long, push into soil/pot, 30cm (1ft) deep and stay at ends of rows with short stakes pushed well into soil; or use 2 horizontal wires just above soil and just below greenhouse roof, running along row, attached at each end to short stakes and bolts in greenhouse structure, tie fillis loosely to each by each plant, or use wire hooks in soil to replace bottom wire; N.B. tomato plants in full crop very heavy,

ensure that greenhouse structure strong and stable. Tie plants to support as they grow, or twist fillis round stems, treat growing tip with care, as brittle; rub off sideshoots in axils of leaves, as soon as they appear, take care not to rub-off growing-tip by mistake; stop plants when required number of trusses present, or when plant has reached limit of space, by breaking off tip of stem just above first leaf above topmost truss; remove any vegetative growth from ends of trusses, and secondary growth from base or low down on plant; as fruit begins to colour on lowest truss, cut off leaves below and up to truss, flush with main stem, to allow light to penetrate and improve air circulation

Harvest pick fruit when full coloured, and swelling has stopped; break off stalk at 'knuckle' by pulling fruit up and towards plant; pick yellow or green fruits at end of season and put in warm, dark place to finish ripening; haulm makes excellent garden compost

Troubles Leaf and stem troubles: red spider mite; caterpillars; whitefly; grey mould; *potato blight*; wilt, soil-borne fungus disease infects through roots, symptoms are wilting in hot sun, yellowing of lower leaves, recovery in evening, followed by premature falling, browning of internal stem tissue near base, remedy: mulch plants close to stem to induce new roots, ventilate well and spray overhead at midday, destroy plants at end of season, plant in different site following year, or use containers/

ring culture; *leafmould* (cladosporium rot), fungus disease as brown velvety patches on leaf underside, yellow on top side in corresponding places, appear on lower leaves first, spreads rapidly, remedy: remove worst affected leaves, ventilate well, do not plant closely, remove leaves to improve air circulation, avoid splashing plants when watering, spray benomyl, use resistant varieties; *stem rot* (didymella), brown sunken areas on stems just above soil level, later on other parts of stem, yellow lower leaves, remedy: if slightly affected pare off with knife and paint captan solution, otherwise destroy plant, and spray remaining plants as precaution, repeat 3 weeks later; *magnesium deficiency*, yellow between veins of older leaves,

Sweet 100 (Cherry tomato)

reduce potash applications or cut out altogether, spray Epsom salts 4 times at weekly intervals 60g/4.5 litres (2oz/gal) water; *leaf rolling*, normal, can indicate wide variation between day and night temperatures. Fruit troubles: *blossom-end rot*, black or grey dry bases to end opposite stalk, irregular water supply, lack of water; *yellow patches or green area* round calyx, too much sun/too little potash, too heavy defoliation can encourage by exposure; *splitting*, irregular water supplies; *fruit sets but does not swell*, dry air at pollination; *flowers fall*, no setting, dry air, lack of soil water; *halo spots* with dark centre on fruit, 'ghost spot', no rot, no treatment, but improve air, watch for grey mould elsewhere

ESSENTIAL FACTS

INDOOR CULTIVATION

Type of plant perennial, grown as annual

Part eaten fruit

In season midsummer–mid autumn; warm areas, early spring onwards

Yield cold house average 3.5kg. (8lb)/plant; heated, 4.5kg (10lb)/plant

Time from sowing to harvest 16 weeks

Size fruit, 1.5–10cm (½–4in) diameter; plants, 30–210cm (1–7ft) tall

Hardiness tender

Seed viable 3 years

Germination period 8–11 days, 16 at low temp.

Culinary uses

There are as many ways to serve tomatoes as there are for onions; their special flavour blends with or enhances that of many other savoury foods and, because of their texture they lend themselves to an enormous variety of recipes. They are certainly among the most versatile of vegetables. Raw, they can be eaten like plums, made into salads consisting only of sliced fresh tomatoes garnished with chopped basil and black pepper, or the classic Italian one of sliced tomato and mozzarella cheese, with the same garnish, or of a mixture of shallots and tomatoes in vinaigrette, a standard French hors d'oeuvre. For mixed salads, the combinations with tomatoes are legion and are limited only by one's imagination. Any meat casserole is improved by the addition of tomatoes; fish dishes à la Provence contain tomatoes, pizzas and quiches will be enriched, and of course they are essential to ratatouille. Home-made tomato soup is delicious, as is home-made tomato sauce, which adds a piquancy and juiciness to all sorts of otherwise rather dry dishes.

Varieties

The average tomato plant is a heavily-leafed one, with clusters of bright yellow flowers (trusses) followed by an average of seven or eight scarlet fruits in a cluster. There are many different variations on the round, bright red fruit commonly seen, many with better flavours, and a considerable choice in size of plant, suitability to temperature and climate, size of fruit and type of growth habit. Plants with yellow, orange or striped yellow-on-red fruit exist; some are plum- or pear-shaped, or with pronounced ridges near the calyx, some cherry size and shape, some even smaller, like strings of large currants. Others are large enough to weigh a pound each. Small plants growing only 1 or 1½ft tall, for pot cultivation on windowsills, are now grown, contrasting the stems or vines of the indoor kinds which are grown in commercial glasshouses to lengths of 4.5m (15ft) or more, by removing all the sideshoots and maintaining the temperature and feeding.

Cultivation

But whether grown indoors or outdoors, all have a need for fertile, well-drained soil, warmth and good light, from seed-sowing until maturity. In cool temperate climates grown outdoors, 4 or 5 clusters are the most that can be expected to ripen. Under glass, up to eight are possible.

Indoor cultivation: Plants grown under glass are usually restricted to a single stem, the cordon method, which is tied to canes or supported by fillis twisted round it. Any side growths are removed, and fruit clusters or trusses allowed to number seven or eight. They can be planted direct into the greenhouse soil, or in containers such as 23cm (9in) pots or grow-bags, or in bottomless rings by ring culture. In this method plants are forced to produce two root systems, one in the ring which absorbs nutrient as well as water, and one in the aggregate beneath the ring – consisting of shingle, fine clinker or similar material – which absorbs water only. This system prevents the build-up of diseases in the growing medium, as the aggregate can be cleaned at the end of every season, and also allows the production of bigger plants. The nutrient-film technique and straw bales are other specialist methods of avoiding this problem.

Outdoor cultivation: Tomatoes can be grown outdoors without any difficulty in warm temperate and tropical climates; they can be sown where they are to grow and start to ripen the fruit naturally early in the summer. In cool temperate regions, they will have to be sown under cover, with artificial heat, and only planted when the night temperature does not drop below 10°C (50°F) or, if necessary, ensuring this with cloche or frame protection. Much shorter plants and fewer trusses will be the norm for cordon-training, but outdoor plants can also be grown as 'bush' tomatoes, which produce a mass of sideshoots, and do not need sideshooting, stopping or staking. It is important to grow a variety suited to outdoor cultivation and to choose a sheltered south-facing position, preferably backed by a wall or fence. Outdoor plants are much more likely to be affected by potato blight.

Super Roma BF

Alicante

CULTIVATION

OUTDOOR
Site and soil sheltered, sunny; well-drained, fertile, alkaline
Soil preparation dig 1–2 spades deep, mix in organic matter in early winter, lime if soil acid in late winter
Sow indoors mid spring, in 18°C (65°F), 2.5cm (1in) apart in tray/pan; 1.5cm (½in) deep
Prick out when 3 leaves present, into 5cm (2in) pots, maintain 16°C (60°F) minimum
Pot into 9cm (3½in) pots
Plant late spring (cloche), or early summer if nights still chilly; 1st truss should be showing
Care space 45 × 75cm (1½ × 2½ft); keep well watered in dry weather; liquid-feed weekly in poor soil
Prune/support treat cordon types as for indoor plants, but stop at 4 trusses, 3 in cold areas, in late midsummer; bush var. allow to grow without check, provide sprawling kind with straw mulch
Harvest break off fruit at knuckle by bending up and inwards; pick remaining yellow/green fruit in early autumn and ripen under cover
Troubles see indoor cultivation

ESSENTIAL FACTS

OUTDOOR CULTIVATION
Type of plant perennial, grown as annual
Part eaten fruit
In season late summer–early autumn
Yield 2.25kg (5lb)/plant
Time from sowing to harvest 20 weeks
Hardiness tender
Seed viable 3 years
Germination period 8–11 days, but often longer

RECIPE
Exotic

Tomato sorbet
*1kg/2lb tomatoes, peeled and
    puréed
2 small apples, peeled, cored and
    cut into small dices
1tsp shallot, minced
1 clove garlic, minced
fresh tarragon leaves, finely
    chopped
1 lemon
100ml/4tbls dry white wine
pinch of curry powder
salt and pepper to taste*
Pour the wine into a saucepan, add the shallots and garlic and cook until the vegetables are soft and the liquid has disappeared. Add the tomatoes, curry powder, juice of the lemon, salt, pepper and the diced apple. Cook gently for about 30 minutes, stirring several times. Leave to cool, then add the tarragon leaves. Pour mixture into a freezer container, cover, seal and freeze for about 1½ hours until mushy. Turn into a bowl and whisk well. Return to the freezer, then freeze for a further 1 hour. Whisk again. Return to freezer and freeze until firm. Transfer to refrigerator 30 minutes before required to soften slightly. Spoon into chilled individual glasses and serve as a starter with a range of crudités or prawns, if desired.

Turnip

(Brassica rapa) CRUCIFERAE

The history of the turnip is very long indeed, stretching back to prehistoric times; it has been a staple part of the diet of northern and central Europeans for many hundreds of years. The Sumerians recorded its use in the Middle East and no stew or soup was without a turnip or two in Anglo-Saxon times. Gerard wrote that it: 'is many times eaten raw, especially of the poore people of Wales, but most commonly boiled.'

The turnip has never had a good culinary press, nor for that matter a good nutritional one, but the maincrop varieties will grow easily and store well through six months of winter, thus supplying some food at a time when it could be in short supply. It replaced potatoes to a large extent, but once these were introduced, the turnip gradually took a minor place on the meal table. However, it continued, and still continues, to be regularly grown, commercially and privately, for human and animal foodstuffs.

Varieties

A plant with such a long history would be certain to have a wide range of varieties, and there are both early and maincrop sorts, typically with white flesh, but some also with yellow, and with green or purple tops. Shape can vary between a flattened globe, round, cylindrical or tankard-shaped and long; in general with white flesh, and often a purple top. The maincrops are excellent for storing, the earlies have to be eaten within a few days of pulling, and are cropped in spring and summer.

Nutritional value

In spite of the general belief that turnips are a watery, tasteless vegetable, good only to provide fibre, they actually have greater nutritional values than the swede, particularly of vitamins, and minerals, and contain quantities of protein and sugars as well as some calories.

Culinary uses

The traditional boiling method of cooking does not do a lot for the flavour, and they are better steamed, then seasoned, and cut into chunks or mashed with butter. They can play a large part in casseroles and soups, and be baked in the oven with a joint, like parsnips. Early turnips, that is those available all through summer, have a delicate flavour quite unlike their winter relatives and the French have devised a number of elegant dishes for them – glazed in sugar and butter and served with lamb or duck or even stuffed and served with triangles of fried bread.

Cultivation

Provided turnips are given the right soil, and this is important, they are no trouble to grow; flea-beetle should be watched for while the plants are young, and water must always be available. Choose the right varieties for the season, and provide firm soil.

CULTIVATION

Site and soil sunny/open; light, well-drained but moist, plenty of humus but manured for previous crop, alkaline

Soil preparation dig 1 spade deep in autumn, lime in winter if necessary; rake in fertilizer 2 weeks before sowing, firm soil by treading

Sow early crop, outdoors, early spring–early summer, rows 23cm (9in) apart, sow thinly; maincrop, outdoors, mid–late summer, rows 30cm (1ft) apart; sow 1.5cm (½in) deep

Thin must thin early, immediately seedlings can be handled, otherwise roots do not swell; thin earlies to 12cm (5in), maincrop to 23cm (9in)

Care watch for flea-beetle while young; keep well watered in dry weather; control weeds, particularly important for turnips

Harvest pull earlies from late spring–late summer, 3cm (1¼in) diameter for raw eating, 6.5cm (2½in) for cooking; dig maincrops from mid-autumn, or when large enough to use, or leave in ground over winter, protect straw/bracken

Store outdoors in clamps, or in sand/dry peat in shed

Troubles flea-beetle; slugs/snails; soil caterpillars

ESSENTIAL FACTS

Type of plant biennial
Part eaten root
In season late spring–late winter; (stored, late autumn–late winter)
Yield roots, earlies, 120–150g (4–5oz) each, maincrop 500g (1¼lb)
Time from sowing to harvest 8–12 weeks
Size roots, 4–10cm (1½–4in)
Hardiness slightly tender
Seed viable 2 years
Germination period 6–14 days

RECIPE
Exotic

Turnips with spinach purée
6 young turnips
chicken or vegetable stock
100g/4oz leaf spinach
salt and pepper
1tbls double cream
circles of bread, at least 1.5cm/½in thick, fried

Peel and cut the tops off the turnips. Decorate edges in a fluted pattern with a sharp knife. Cook in boiling stock until tender but still firm, about 10–15 minutes. Make a shallow well in the turnip with a spoon, reserving the scooped-out turnip flesh. Keep turnips warm. Remove spinach from stalks, wash well and sweat over low heat until soft. Purée the spinach. Bring the cream to the boil, add the reserved turnip flesh and spinach, purée and season. Fill each turnip with purée and serve on circles of fried bread or as an accompaniment to roasted lamb or duck.

Exotic and unusual vegetables

While the assortment of commonly grown vegetables is considerable, there are a good many more rarely seen or grown, which are worth making part of the everyday diet because of their interesting flavours and nutritive content. Some are easily grown in cool-temperate climates, some will only survive in such climates with protection because they are native to warm-temperate or sub-tropical zones, and others come from the tropics, to be obtained only from stores in temperate zones. The descriptive list which follows supplies details of their appearance, flavour, uses, and cultivation where practicable, in the hope that readers will be encouraged to experiment with an even wider assortment of vegetables, and so expand their knowledge of an important part of a healthy diet.

YAM
(*Dioscorea spp.*, DIOSCOREACEAE)

Perennial, mostly tropical zones; *D. batatas* is said to be hardy in cool-temperate zones but not grown as tuber roots very deep down in the soil and difficult to obtain. Part eaten is root tuber, long and rectangular, with brown skin and white or orange flesh; size variable, from 30–90cm (12–36in) but can be much larger and weigh 40kg (100lb). Plants produce either a single large yam, or a cluster of smaller ones better flavoured and rather sweet. Widely eaten all over tropics, and in Central America and Brazil but contain mostly starch and little protein. Cooked by boiling and mashing, peel before cooking, also roast, fry, casserole, and curry them. Twining plant to 3m (10ft), often with handsome leaves and insignificant flowers. Needs very rich deep soil, dug 60cm (2ft) deep; plant 120 × 90cm (48 × 36in) apart in early spring; use pieces of tuber each with two or three eyes. Mature about nine months later, dig up and store or leave in ground until required. ▽

CARDOON
(*Cynara cardunculus*, COMPOSITAE)

Cool temperate to subtropical zone. Slightly tender perennial grown as annual. Forms attractive large plant to 180cm (6ft) tall if allowed to grow to full size, with large prickly leaves like globe artichoke leaves, only more prickly, and similar purple flowerheads. Widely grown in France and Italy for stems which are blanched at end of growing season, like celery. Flavour is different to most vegetables, rather like a combination of celery and globe artichoke. Stems are crisp but must be cooked before eating, removing the stringy outer parts first. If boiling, allow at least half an hour. Cooked cardoons can be served tossed in butter, folded into a Béchamel or Mornay sauce, added to soups or stews or used cold for salads. Station sow seed outdoors in mid spring in cool–temperate climates, spaced 45cm (1½ft) apart, 90cm (3ft) between rows, in deep moist fertile soil and a sunny place. Allow one plant at each position, water well throughout the summer, especially in hot dry weather, tie to a stake and mulch in midsummer. Start blanching early in late summer as for celery, and dig up whole plant about eight weeks later; ensure that leaves are dry and free from slugs before starting to blanch. ▽

CHAYOTE
(Choko, Cristophine, Cayote, *Sechium edule*, CUCURBITACEAE)

Perennial climbing plant, tropical and sub-tropical zones. Pear-shaped, 10–15cm (4–6in) long, pale green or white, weight for eating about 1kg (2lb). Delicate flavour, texture like marrow but firm when cooked. Treat like courgette or marrow, boil, steam, bake stuffed with meat or seafood mixtures, or stir-fry. Plant grows to 3.6m (12ft) and has yellow flowers, fruits a few months after planting, contains one seed. Usually grown by planting whole fruit in mound of rich soil and then treating like marrow, supplying plenty of water and plant food. Later-sown plants will crop in spring. ▽

TRUFFLES
(Black, White, *Pachyphloeus melanoxanthus* (*Tuberosum melanosporum*), *Choiromyces meandriformis* (*T. magnatum*), TUBERALES) ▷

Truffles are simply a form of wild mushroom which grows beneath the soil surface, attached to the roots of oaks, beech, poplar, hazel and willow. They are highly prized because they are rare and are picked in France and Italy with the help of truffle hounds and specially trained pigs. They are irregularly round, with a warty surface, varying in size between a walnut and a potato. Black or white truffles are either black on the outside with white veining of the flesh, or yellowish to beige with similarly coloured, white-veined flesh. The black truffle, often known as the Périgord truffle, is the most well-known and has a unique aroma and taste. Fresh or less successfully canned, this truffle is extremely expensive. However, a small amount will pervade an entire dish. The white truffle, found in Italy, has a slightly less powerful odour but is also expensive. Truffles can be added to many savoury dishes and are often included in *paté de foie gras*. A little added to risotto, fondues, eggs and pasta will transform the dish into a feast. The white truffle is also particularly good eaten raw, sliced thinly and added to salads. Truffles should normally not be washed or peeled, but if washing is essential, use dry white wine.

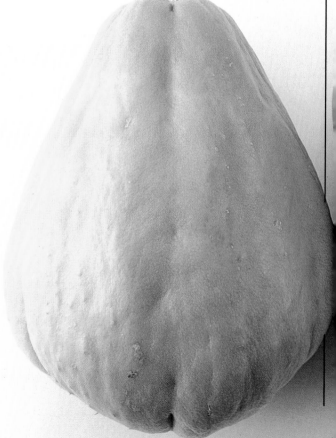

KALE
(Borecole, *Brassica oleracea* 'Acephala', CRUCIFERAE)

Biennial, cool- to warm-temperate zones. Leaves are the part eaten; many cultivars, best is 'Pentland Brig' which has edible flowering shoots like sprouting broccoli as well, following leaves; harvest late autumn–mid-spring. Cook as rest of cabbage family. Very hardy plant, stands severe frost; grows about 60cm (24in) tall. Sow mid–late spring in nursery bed, thin to 7cm (3in) apart, transplant when about 12cm (5in) tall, and space 45cm (18in) apart each way. Cook and serve blanched leaf shoots as for celery, leaves can be used in salads.

OYSTER MUSHROOM
(*Pleurotus ostreatus,* AGARICINEAE)

The oyster mushroom is so-called because of the shape of its cap; this is deep blue-grey, almost black, when young, changing to brown or fawn as it ages. The gills beneath are white and widely spaced, attached to the short stem which in turn grows from the trunks of such trees as beech, poplar, mulberry and other broad-leaved species, also tree stumps, fence posts and occasionally conifers. Its presence is unfortunately an indication that the host tree is dying or is likely to, as the fungus by then is present throughout the wood, and the oyster mushroom is the fruiting body producing spores with which to infect other host trees. It appears in autumn and winter. The flavour is delicate, and the flesh has a rather chewy texture; when gently sautéed or stir-fried with other ingredients, as is popular in Chinese cooking, it adds a distinctive flavour and texture to any dish. Oyster mushrooms dry well, and can easily be cultivated by seeding spores on to logs of suitable woods. Indeed, they are now cultivated commercially in many countries.

HAMBURG PARSLEY
(Turnip-rooted parsley, *Petroselinum crispum* 'Tuberosum', UMBELLIFERAE)

Biennial, cool- to warm-temperate zones. Grown for its fleshy root, creamy white, about 15 × 7.5cm (6 × 3in), flavour of celery/parsnip; leaves can be used like parsley. Popular in German and Eastern European cooking, where it is frequently used to add flavour to soups and stews. Root can be boiled or steamed whole, or sliced or grated raw and used in salads with dressings. Plant about 15cm (6in) tall, flowering stem 30–45cm (12–18in). Sow seed outdoors 1.5cm (½in) deep in mid–late spring, in sun/slight shade and average-moist soil, thin to 23cm (9in) apart each way, water well in drought and dig up in autumn or as required.

MOOLI
(Icicle Radish, Rettich, *Raphanus sativus*, CRUCIFERAE)

A mild-flavoured radish similar in shape to the Continental white radish. For cultivation details see *Radish* p. 114. Use thinly sliced in salads or cook briefly in boiling salted water and toss in butter.

SWEET POTATO
(Kumara, batata, *Ipomoea batatas*, CONVOLVULACEAE)

Perennial, subtropic to tropical zones (warm-temperate with protection). Root tubers edible, white or yellow flesh, purple or white skins, elongated or roundish, large, up to 23 × 10cm (9 × 4in); contain much starch and have a sweetish, chestnut flavour. Used extensively in the US, particularly the south, where they are commonly served glazed with Thanksgiving turkey. Bake and use with richly flavoured meat and game, or boil and then mash; flour is used in pastry industry. Can also be made into dessert soufflés and cakes. Plant is climbing with heart-shaped leaves and purple flowers. Grown from stem cuttings in tropics, from forced shoots rooted in winter and planted outdoors in spring, in warm-temperate zones. Needs five months from planting out to mature; light rich soil and sun, plant 45 × 90cm (18 × 36in) apart on ridge; 20 plants sufficient for average family. Dig up when plant completely yellow and treat as potatoes, store for winter.

TARO
(Eddoe, kandalia, dasheen; *Colocasia esculenta*, ARACEAE)

Perennial, subtropics to tropical zones. Grown for its tubers, similar to potatoes in size and shape, skin is thick and fibrous, light brown; high starch content, more protein and stronger, better flavour than European potato; popular in West Indian and Caribbean cooking; cook as potatoes, but for longer, as they contain less water, flavour with grated nutmeg. Plants have shield-shaped leaves, height about 120cm (4ft), flower-head a white spathe and spadix. Need moist soil and sun, plant in spring 90 × 90cm (36 × 36in) apart and dig about six months later, three with some varieties.

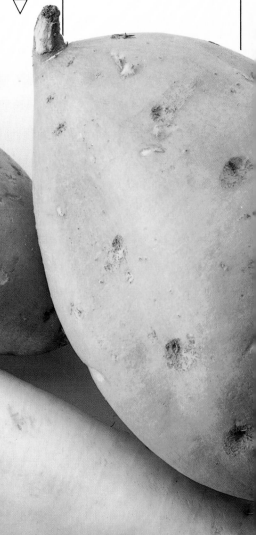

CEPS
(Cèpes, Boletus, *B. edulis,*
BOLETINEAE)

Ceps are rather stout wild mushrooms, and this species has a brown shining cap with white flesh, beneath which is a spongy mass of vertical tubes, rather than gills, which are white to start with, later a yellow-green colour, from which the spores are produced. Ceps can be 10cm (4in) high, with stalks about 4cm (1¼in) thick, but may be twice as large. They are found in late summer and autumn in woodland consisting of beech, oak, chestnut or conifers, where they may be found associating with the trees' roots, as do truffles; also, like them, they cannot be artificially cultivated. They have a nutty, distinctive flavour, and in Europe are frequently added to soups and stews or served, braised, with meat and fish dishes. One popular way of serving them is *à la bordelaise,* sautéed in olive oil with parsley and garlic. Ceps are often available canned or dried; if dried, they should be soaked in a little water before using.

ASPARAGUS PEA
(*Tetragonolobus purpureus,*
LEGUMINOSAE)

Cool-temperate zone. Small green pods about 2.5cm (1in) long, eaten whole; flavour similar to asparagus. Steam/boil or stir-fry and eat hot. Tough and stringy if longer than 2.5cm (1in). An annual, trailing to about 45cm (18in), has deep red-brown pea flowers from midsummer. Sow seed mid spring 2.5cm (1in) deep in seed trays, space 10cm (4in) apart, and keep at about 16°C (60°F). Plant out late in late spring 38 × 60cm (15 × 24in) in sun and averagely good soil. Use canes and wires to make growth vertical.

CHILLIS
(Hot peppers, *Capsicum frutescens,*
SOLANACEAE)

Perennial grown as annual, warm-temperate to tropical zones (cool-temperate zone with protection against frost). Small fruits varied in size from long to triangular, usually cultivated as long, narrow and pointed, 10 × 1.5cm (4 × ½in) long, green turning red, extremely hot flavour. Use fresh (shredded) or dried, removal of seeds and blanching takes away much of hotness; used extensively in Mexican cooking and as a spice in Indian cuisine. Add to meat dishes such as casseroles or curries, do not eat raw. Plant similar to sweet pepper in size and appearance. Sow seed in temperature of 18°C (65°F); in containers supply 23cm (9in) final pot size, outdoors space 30cm (12in) apart, each way. Plant in sun and rich soil, and water well.

PAK CHOI
(Brassica sinensis CRUCIFERAE)

Cool temperate to subtropical zone. Pak Choi is a collective name for a group of leafy lettuce-like brassicas, very popular in Chinese cuisine, grown as annuals from seed with rounded, entire or toothed-edge leaves and thick white midribs and stems. Both parts can be eaten, raw or cooked, and the crisp white midribs are particularly succulent. Very useful as they will provide fresh salads in winter from a cold greenhouse or frame in cool-temperate climates. In Chinese cooking, mostly stir-fried with a little soy or mushroom sauce added or lightly braised. Sow seed outdoors in spring for summer cropping, or in late summer for cropping under cover, or in pots for greenhouse planting in autumn. Space about 25cm (10in) apart and keep summer plants well watered and slightly shaded to prevent bolting. Varieties include 'Chinese Pak Choi', cold-resistant and slightly sweet flavour; 'Shanghai Pak Choi'; 'Japanese white celery mustard' grown in large quantities in northern Japan and 'Purple Pak Choi' (Hon Tsai Tsai), cultivated for its purple flower stems which can be stir-fried or served with sauces.

▽
bottom

SALSIFY
(Vegetable oyster, *Tragopogon porrifolius,* COMPOSITAE) ▷

Perennial grown as annual/biennial, cool- to warm-temperate zones. White root is edible, long and tapering, slight flavour of oysters; popular in France and Italy, it can be baked, steamed or boiled, and served alone or with a white or cheese sauce, fried in batter, puréed or made into soup. Available mid-autumn–late winter. Forms rosette of long, narrow grass-like leaves, with purple flower on 60–90cm (24–36in) stem. Sow seed 1.5cm (½in) deep in mid–late spring in rows 30cm (12in) apart, thin to 25cm (10in). Use average soil and sun/slight shade. Water well in drought. Dig as required, finish by end of early winter. Use new shoots in spring like asparagus, cut at 10–15cm (4–6in) tall. Also blanch in early winter with 15cm (6in) soil above crown and cut shoots early spring.

SCORZONERA
(Black salsify, *Scorzonera hispanica,* COMPOSITAE)

Perennial, grown as annual/biennial. Cool- to warm-temperate zones. Fleshy root is edible, has black skin and white flesh, contains inulin, similar in flavour to salsify but slightly sweeter and more delicate; length about 30cm (12in) by 2.5cm (1in) thick. Cook by steaming, boiling or braising, then rub off skin; serve hot or cold in salads, fried in butter or with sauces. Plant has rosette of broader leaves than salsify at ground-level, and yellow flowers. Grow as for salsify, but also sow in early summer.

▽

ARTICHOKE, CHINESE △
(Crosnes, *Stachys affinis,* LABIATAE)

Cool-temperate zone. Small, spirally-shaped, cream-coloured tubers, 2.5–7.5cm (1–3in) long, 1.5cm (½in) thick; delicate flavour similar to salsify root and much appreciated in France where they are widely grown. Plant grows 15–23cm (6–9in) tall with lilac flowers at midsummer. Use as a side vegetable, unpeeled, boil/steam for 10 minutes, then toss in butter, serve with sauce, or lemon, or use in a salad. Plant early spring 7cm (3in) deep, in well-drained, fertile soil and a sunny place, space 23 × 38cm (9 × 15in) apart. Harvest late autumn–late winter, dig on day to be eaten.

◁
KOHLRABI
(Turnip-rooted cabbage, *Brassica oleracea* 'Gongyloides' CRUCIFERAE)

Biennial, cool- to warm-temperate zones. Swollen base of stem shaped like turnip from which leaves come is part eaten, best when small, between 4 and 7cm (1½ and 3in) diameter, otherwise tough and flavourless; purple and green cultivars. It is a popular vegetable in Eastern Europe and can be used to replace turnips or celeriac in a recipe. Boil or steam whole if small, otherwise slice first; also use raw, grated or chopped for salads. Plant about 23–30cm (9–12in) tall. Sow seed mid spring–midsummer, also midsummer but may bolt, and thin to 23cm (9in) apart each way; grow fast for most tender roots, ready in two months. Supply sun, fertile soil and plenty of water. Leave in soil to early winter, but only the later sowings made in midsummer.

ASPARAGUS LETTUCE ▷
(Celtuce, stem lettuce, *Lactuca angustana*, COMPOSITAE)

Cool temperate to subtropical zone. Annual grown from seed for its stems and leaves, looks like very elongated cos lettuce, with long white midribs and narrow leaves forming a central stem which has a delicate flavour and crunchy texture; can be sliced and eaten raw, or cooked like celery or Swiss chard; leaves treated as lettuce, and used mainly in salads. Sow and cultivate as lettuce; sow seed outdoors in spring and thin to about 25cm (10in); plenty of water essential to prevent bolting. Use leaves when young and harvest stems when plant about 30cm (1ft) tall.

Salad leaves and sprouts

Many leafy plants, some of them wild and regarded as weeds, make excellent salad leaves; also, a great many can be easily cultivated to give the gardener a variety of salad ingredients all the year round. All the leaves listed here can be eaten in their raw state and most are rich in vitamin C and minerals. Sprouting seeds are particularly versatile, since they have the advantage of year-round, indoor cultivation. Young leaves and sprouts, used imaginatively, can add infinite variety, not only to the vegetarian table but to the family meal and even the gourmet dinner.

WATERCRESS ▷
(*Nasturtium officinale*, CRUCIFERAE)

Cool temperate to tropical. Hardy perennial growing in water either floating or rooted, with dark green, shiny leaves, pinnate when mature, and fleshy stems. Low-growing to about 10cm (4in), sprawling habit, minute white flowers in summer. Leaves and stems have hot, peppery flavour; many uses including salads, garnishes, sauces and as delicious summer soup. Watercress makes a fine salad either on its own or mixed with other leaves. It also combines well in a salad with mushrooms, oranges and pears. Contains range of vitamins including A, B, C, D and E, and various minerals, eg. iron, manganese, iodine, phosphorus and calcium; has considerable medicinal properties. Root stem cuttings in water – shop-bought watercress will do this – and grow along edges of a stream, or plant in garden soil, as moist as possible and keep well watered, or grow in well-drained pots standing in water frequently changed; supply a little shade.

ROCKET
(Arugula, roquette, Italian cress, *Eruca vesicaria,* CRUCIFERAE)

Cool temperate to tropical zone. Annual, grows rapidly from seed to between 20 and 100cm (8–40in) depending on soil and climate. Many rosette and stem leaves, deeply lobed; yellow flowers late summer. It has a delicious, slightly hot, savoury flavour, best in young leaves, quite unlike other salad leaves, and adding most unusual flavour to the average salad. Grows easily from seed, sown outdoors spring (late spring for cool–temperate climates) and does best in warm, dry, sunny conditions; most soils suitable. Late summer sowings will survive a little frost, but for winter use grow under cover or indoors. ▽

Watercress, avocado and mushroom salad with blue cheese dressing.

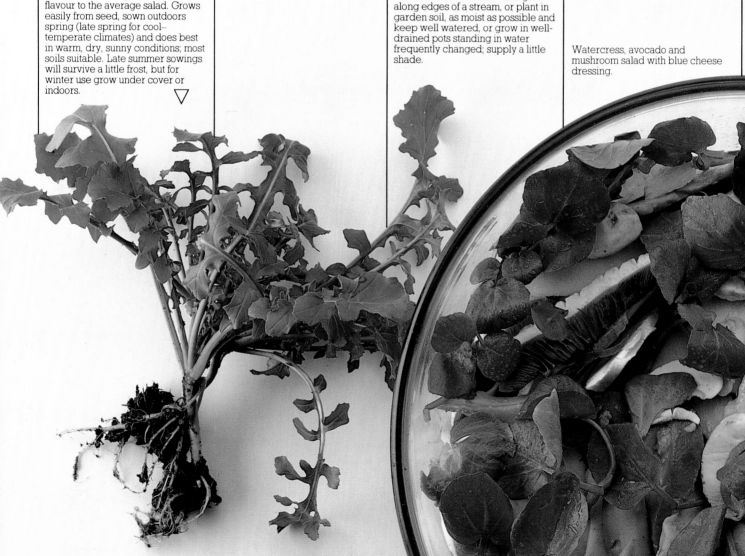

LANDCRESS △
(Winter cress, American cress,
Barbarea verna, CRUCIFERAE)

Cool temperate zone. Dark green,
shiny leaves very similar to
watercress, for which it can be
substituted. It is an excellent winter
salad ingredient, though it is also
available in summer and best used
in a mixed salad with endive,
chicory or celery, for example; good
for garnishes and as a soup. The
flavour is hot and peppery. A
biennial, height in flower is about
38cm (15in), otherwise low-growing.
For winter picking, sow mid–late
summer in moist soil and a slightly
shaded place, space 15cm (6in)
apart each way. Also sow in spring
for summer crops. Flea beetle can
devastate seedlings and young
plants.

PURSLANE △
(*Portulaca oleracea,* PORTULACEAE)

Cool temperate to tropical zone.
Annual, grows rapidly from seed,
has fleshy, red-tinged stems and
thick spoon-shaped shiny leaves.
Height about 30cm (1ft) bushy and
somewhat sprawling, with tiny
yellow flowers in mid–late summer.
It contains considerable vitamin C
and was once used as a remedy for
scurvy; is also a tonic and has
considerable medicinal use. The
leaves supply a cooling flavour and
are very juicy, so are particularly
good for summer salads. Sow
outdoors in a sunny place and well-
drained soil, in rows in late spring–
early summer, and thin to 15–20cm
(6–8in). In cool dull weather, it
grows poorly. Can be used about
six–eight weeks after sowing; use
leaves when young and keep
picking to maintain supply.

NASTURTIUM △
(*Tropaeolum majus,*
TROPAEOLACEAE)

Warm temperate to tropical zone.
Ornamental climbing and dwarf
annual with bright orange, orange-
red or brown-red flowers produced
in profusion throughout summer,
and shield-shaped dark green or
white-variegated leaves which have
a peppery flavour. Like dandelion,
only very young leaves should be
used in salads and they are
especially good with cucumber.
The seeds are also used, as
substitute capers, and the flowers in
salads; they have less flavour than
the leaves but are quite edible and

have considerable decorative value.
Sow seed outdoors in spring or, in
cool temperate climates, late spring,
in any average soil – more leaves
are produced in moist rich soils,
most flowers in poor dry ones – and
a sunny place. Allow 15cm (6in)
spacing and supply supports for
climbers, or allow to grow as
groundcover. Will self-seed.

MUSTARD AND CRESS
(*Brassica alba, Lepidium sativum,*
CRUCIFERAE)

The seedlings of these two plants
are used mixed for salads,
sandwiches and as garnishes; both
have a hot flavour. Nowadays, much
shop-bought mustard and cress is, in
fact, seedlings of rape, so home-
growing – which is so simple – is
necessary to get the true mixture.
Mustard germinates more quickly
than cress, so should be sown three
days after cress; allow a week to
cutting. Use the seeds in equal
quantities and sow on moist peat in
cartons or shallow trays, together or
separately. Keep the peat moist and
cover the container until
germination starts; keep in about
16°C (60°F). Sow at regular intervals
to ensure a constant supply. It can
also be grown outdoors but is easily
splashed with mud; for winter
cutting, in this case, sow in a frame
or under cloches. ▽

bottom

raddichio

SORREL
(*Rumex acetosa* (English), *R. scutatus* (French), POLYGONACEAE) △

Two sorrels, English and French, are widely used in salads, but the French variety has the better flavour – rather like young spinach leaves subtly flavoured with lemon. English sorrel: cool temperate to subtropical zone. Strong-growing perennial with tough vertical flower stem to 60–120cm (2–4ft) tall, and arrow-head shaped leaves up to 20cm (8in) long, sharply acid-flavoured; flowers are tiny, reddish, in a spike in summer. Root penetrates deeply. Leaves contain oxalic and tartaric acid. Easily grown from seed sown outdoors in spring, thinned to 30cm (1ft) apart, or from pieces of root, planted in autumn. French sorrel: warm temperate to tropical. Bushy low-growing plant, with sprawling stems and grey-green fleshy, shield-shaped leaves about 4 × 4cm (1½ × 1½in); flowers green, tinted reddish in midsummer, in loose clusters on stems 45cm (1½ft) tall. Leaves pleasant sharp lemon flavour, especially when young and widely used in Europe in soups and sauces, as well as salads. Sow outdoors late spring in well-drained soil, and sun or a little shade, and thin to 30cm (1ft) each way; will self-sow in moderation. Protect in severe winters.

LAMB'S LETTUCE
(Corn salad, Mâche *Valerianella locusta*, VALERIANACEAE) ▷

Cool temperate zone. Very hardy, small annual plant forming a rosette of spoon-shaped leaves; the whole plant is about 12–15cm (5–6in) wide. It is available all year, but most popular in winter and early spring when few plants can supply fresh green leaves. It is mildly flavoured, thin and crisp when fresh. In France where it is widely popular, it is frequently eaten in a salad with beetroot. Sow seed early–late summer for winter–spring cropping, space 10–15cm (4–6in) apart each way and make sure the soil remains moist from sowing time, otherwise germination will be poor. Protect in winter for better quality plants; most soils suitable.

'SALAD BOWL' LETTUCE
(*Lactuca sativa*, COMPOSITAE) △

Cool temperate to subtropical zone. Annual plants grown from seed, these lettuce varieties are grown to provide single leaves rather than a cluster forming the tight hearts of the familiar cabbage or butterhead varieties. The leaves can be stripped off singly throughout growing season from time when plants are half grown, so this is a very practical lettuce for the salad lover to grow. None form a heart. They have curly or wavy margins to the leaves, which may be oak-leaf or tinted red. Try this type of lettuce with some American-style dressings such as Thousand Islands or Blue Cheese or with a vinaigrette made with Seville orange juice instead of lemon or vinegar. Grow these lettuces as for summer lettuce, and space about 15cm (6in) apart in rows, or sow broadcast and thin to 10cm (4in). One or two sowings should be sufficient to provide leaves all through summer. They are not winter lettuces, but do germinate better in high temperatures than hearted lettuce.

DANDELION
(*Taraxacum officinale*, COMPOSITAE) △

Cool temperate zone. A well-known perennial plant, regarded as a weed, with long, deeply serrated leaves and yellow flowers followed by a fluffy round 'clock' or seedhead. It spreads rapidly by seed, and can also be propagated by pieces of root. The leaves have a sharp, bitter taste, making them a good partner for purslane, and are best eaten when very young, or can be blanched. The cultivated variety is less bitter, with thick leaves. Dandelion is said to have a good effect on the digestion and liver and is sold in large quantities in France where *pissenlit au lard* – a salad of young dandelion leaves and warm pieces of salt pork or bacon with a vinaigrette dressing – is very popular.

SALAD BURNET
(*Poterium sanguisorba*, ROSACEAE) △

Cool temperate to subtropical. Hardy, low-growing perennial with a woody root-stock; attractive feathery leaves coloured blue- to grey-green in loose clusters mostly at soil level, but also on the flower stems. Is a good edging to ornamental beds and borders, and makes a good groundcover. Tiny nondescript green or red-tinged flowers in ball-like head appear in mid–late summer. The leaves have a pronounced cucumber flavour and contain vitamin C; they wilt quickly so are best added to salad immediately before serving. They also add a delicate flavour to sauces, summer wine punch, and fruit cups.

SEED SPROUTS

Can be obtained from a variety of plant seeds such as fenugreek (pungent smelling), mung, soya and adzuki beans, alfalfa, lentil, mustard, cress, chick peas, radish, rye, wheat and triticale, a hybrid between the last two. All should be tasty. Mung sprouts are the Chinese bean sprouts so often seen, adzuki beans are the Japanese version. All have a high nutritive value as regards protein, vitamins and minerals, and contain unsaturated oils. Easily grown at any time of year; sow seeds evenly on shallow layer of moist peat or other sterile absorbent material in tray, cover with blown-up polythene bag and put in dark in warm place, such as airing-cupboard (do not forget!). Because they are so nutritious and easy to grow, they make excellent salads at any time of the year. Keep moist and use when at height required; for green seed-leaves, place in light for 1–2 days after germination. Use raw alone or in salads after washing, or stir-fry – bean sprouts especially good this way.

▷

Sow seed outdoors in spring and allow 20cm (8in) between plants which do best in chalky soil and will self-sow. Leaves remain all winter, especially if plants cloched.

Beansprout salad with carrots, peppers and peanuts with sesame oil and soy dressing.

Varieties

The varieties of each vegetable listed here provide a good representative range of those available in Great Britain. For more unusual varieties, it is worth trying some of the specialist seed and plant suppliers listed on page 156. Also seed catalogues each year contain new varieties of common vegetables and you may discover many which are particularly suited to your garden.

Red Drumhead Cabbage

Musselburgh Leeks

(available as seed, unless otherwise stated)

Artichoke, globe
'Green Globe'

Artichoke, Jerusalem
Only available as such; tubers

Asparagus
'Martha Washington', standard var., 'Connover's Colossal', large; as crowns: 1-yr-old 'Larac', early; 1-yr-old 'Lorella'; 2-yr-old 'Suttons Perfection'

Aubergine
'Black Prince', F1 hybrid, outdoor; 'Claresse', early; 'Dusky', very early, TMV resistant, F1 hybrid; 'Oriental Eggplant', white fruit maturing yellow, spicy flavour

Bean, broad
'Aquadulce Claudia', sow late autumn; 'Express', earliest cropper; 'Imperial White', white seeds, very long pods; 'Masterpiece Green Longpod', green seeds, good flavour; 'The Sutton', dwarf to 30cm (12in), good under cloche from autumn sowing

Bean, French
Climbing 'Blue Lake', also for haricot beans; Climbing 'Purple-podded', cooks green; 'Chevrier Vert', use also as haricots flageolets or haricots secs; 'Kinghorn Wax', yellow pods; 'Tendergreen', early; 'The Prince', heavy, long cropping

Bean, runner
'Hammonds Dwarf', like bush bean, dwarf; 'Painted Lady', red and white flowers; 'Red Knight', red flowers; 'Scarlet Emperor', early; 'Streamline', very long pods; 'Sunset', salmon-coloured flowers, less likely to be attacked by

sparrows; 'White Achievement', white flowers, ditto

Beetroot
'Boltardy', round, early/maincrop, resistant to bolting; 'Burpees Golden', globe, yellow; 'Cheltenham Greentop', tapering, maincrop; 'Detroit Crimson Globe', all-purpose; 'Forono', cylindrical, good colour, uniform shape; 'Monopoly', single-seeded, globe

Brussels sprouts
'Bedford Winter Harvest', crop mid-autumn to late winter; 'Fortress', midwinter-early spring; 'Noisette', small sprouts, excellent nutty flavour; 'Roodnerf', late autumn to end early winter; 'Peer Gynt', dwarf habit, heavy even cropper; mid autumn to end early winter; 'Rubine', red sprouts; 'Widgeon', late, from early winter on, resistant mildew, good flavour

Cabbage
Summer-maturing: 'Greyhound', pointed; 'Hispi', F1 hybrid, pointed; autumn/winter: 'Golden Acre', round, late summer-autumn; 'Winningstadt', pointed, late summer/autumn; 'Minicole', round, small, green to white, early autumn/early winter; 'Red Drumhead', red cabbage up to end of early winter; 'Ruby Ball', F1 hybrid, small, red-tinged, savoy, mid-winter; 'Holland Late White', white, late autumn-late winter, for salad/coleslaw; spring: 'Harbinger', pointed, from early spring; 'Offenham Flower of Spring', pointed, mid-late spring; Chinese: 'Pe Tsai', sow midsummer; 'Two Seasons', sow spring-midsummer, bolt-resistant

Calabrese
'Green Comet', large, early; 'Corvet', green, late; 'Romanesco', white, conical, autumn/early winter

Carrot
'Amsterdam Forcing', very early, cylindrical; 'Autumn King', cylindrical, maincrop for storage; 'Early French Frame', round, for forcing and succession; 'Favourite', stump-rooted, maincrop; 'Royal Chantenay', stump-rooted, early maincrop

Cauliflower
Summer-cropping: 'Alpha Polaris', quick-maturing; 'All the year round', also early autumn and late spring; autumn-winter: 'Autumn Giant-Autumn Protecting', late autumn-early winter; 'Barrier Reef', mid autumn, medium size; 'Wallaby', early-mid autumn; spring: 'English Winter-Reading Giant', mid spring; 'Northern Star', late spring; 'Purple Cape', purple heads, late winter-early spring, good flavour

Celeriac
'Tellus', quick-maturing; 'Giant Prague', large

Celery
Winter-cropping: 'Giant White', 'Giant Pink', 'American Green'; 'Giant Red'; summer-cropping: 'Golden Self-blanching', white; 'Lathon Self-blanching', more resistant to bolting

Chicory
'Sugarloaf' (Pain de Sucre), autumn-early winter; 'Witloof' (Brussels Chicory), blanch for winter; 'Snowflake', all winter, hardy outdoors; red chicory: 'Rossa di Treviso', like cos, flushed red in

winter; 'Rossa di Verona', round, red in winter; 'Variegata di Chioggia', round, red and white variegated in winter

Cucumber
Indoor: 'Butchers Disease-resisting', 'Femspot', all-female flowers; ridge Outdoor: 'Burpless Tasty Green', F1 hybrid, 23cm (9in) long fruits; 'Bush Crop', F1 hybrid, bush habit, not trailing; 'Crystal Apple', round pale green fruits; 'Hokus', gherkin, for pickling; 'Patiopik', F1 hybrid, small, early cropping

Endive
'Moss Curled'; 'Batavian Green'

Fennel
'Perfection', round, less likely to bolt; 'Sirio', large, quick-maturing bulbs

Leek
'Giant Winter-Catalina', new variety, long-standing; 'The Lyon-Prizetaker', excellent quality; 'Musselburgh', general use, any conditions

Lettuce
Summer-cropping: Cabbage – 'All the year round', slow to bolt; 'Sigmaball', large, resistant to tip-burn; 'Tom Thumb', small, early-maturing; 'Avon defiance', mildew-resistant; 'Continuity', red-tinged; Crisphead – 'Lake Nyah', good in midsummer; 'Webbs Wonderful', large, heavy, crunchy and nutty; Cut-and-come-again – 'Salad Bowl', no heart, wavy-edged leaves; 'Red Salad Bowl', red-tinted leaves; Cos-'Barcarolle', large, dark green; 'Little Gem', small, good flavour; winter-cropping (with greenhouse protection): 'Dandie', can crop late autumn early spring; 'Kwiek', late

Squash, Table Ace

Seakale Beet

autumn-early winter; both cabbage types; spring cropping: 'Arctic King', cabbage; 'May Queen', cabbage, red-tinted; 'Lobjoit's Green', large dark green; 'Valdor', cabbage, large, resistant to cold, wet; 'Winter Density', cross between cos and cabbage, medium size, delicious nutty flavour, very hardy, all need cloche/tunnel protection

Marrow & Courgette
Bush: 'Tender and True', early; 'Custard Yellow', plate-shaped, scalloped edges, 20 × 4cm (8 × 1½in), yellow; trailing: 'Long Green Striped'; Vegetable Spaghetti', flesh in long strands, scoop out when cooked; Courgette/zucchini: 'Burpee Golden Zucchini', bush; 'Green Bush F1 hybrid', early

Okra
'Long Green', 'Pentagreen', 'Green Velvet', all similar, strong growing and cropping heavily

Onion
'Bedfordshire Champion', round; 'Giant Fen Globe', round; 'Noordhollandse Bloedrede', round, pink; 'White Spanish', flat, large, mild flavour; Japanese: 'Extra Early Kaizuka'; pickling: 'Quick-silver'; shallots: 'Giant Long-keeping Red'; 'Hative de Niort'; spring: 'White Lisbon'

Parsnip
'Avonresister', short, resistant to canker; 'Offenham', good on shallow soil; 'Tender and True', long, good exhibition

Pea
Early: 'Feltham First', 45cm (18in), sow autumn; 'Kelvedon Wonder', 45cm (18in), sow spring; main-crop:

'Onward', 60cm (24in); 'Senator', 75cm (30in), mange-tout: 'Oregon Sugarpod', 90-120cm (36-48in); 'Sugarsnap Eat-all', 150cm (60in); petit-pois: 'Waverex', 45cm (18in); 'Purple-podded', 150cm (60in)

Pepper, sweet
'Big Bertha', very large fruit; 'Canapé', very hardy, early; 'Early Prolific', early, heavy cropper; 'Triton' small, 25cm (10in), containers: 'Yellow Lantern', yellow when ripe

Potato
Early: 'Foremost', high yeild; 'Maris Bard', very early, good flavour; 'Pentland Javelin', latish, scab resistant; maincrop: 'Desirée, red skin, heavy cropper, very good flavour, drought-resistant; 'Drayton', red, like King Edward but higher yield and more disease-resistance; 'Golden Wonder', yellow flesh, excellent flavour, needs good conditions; 'Maris Piper', heavy yield, cooks well; 'Pink Fir Apple', red skin, long tubers, yellow flesh, light yield, but excellent flavour, good for salads

Pumpkin & Squash
'Hubbard Squash', warty orange fruit, pear-shape, about 25cm (10in) long; 'Hundredweight', yellow or orange skin, very large, round; 'Spirit', F1 hybrid, pale green, globe, about 30cm (12in) long; 'Table Ace', bush, orange flesh; 'Melon Squash', green skin, long curved, sweetest after storage

Radish
Summer: 'Saxerre', early, red, round; 'Scarlet Globe', round, general purpose; 'Sparkler', red

with white tips, round; 'French Breakfast Crimson'; long, red with white tips; 'Minowase Summer No 2', white, long, late summer and autumn; Winter: 'Black Spanish Round', black skin, white flesh; 'China Rose', rose skin, white flesh, cylindrical; 'Violet de Gournay', purple

Rhubarb
'Glaskins Perpetual', green, pull all summer, seed; 'Timperley Early', green, earliest; 'Victoria', red, late, thick stems; 'Prince Albert', red, good flavour

Spinach
'Broad-leaved Prickly', winter crop; 'Longstanding Round', summer crop; Perpetual spinach, large leaves, summer-winter; New Zealand, summer-autumn, trailing

Sprouting broccoli
'Purple Sprouting', purple florets early-mid spring; 'White Sprouting', white florets, mid-late spring; there are early and late strains of both these; 'Nine-star Perennial', white florets, early-mid spring, permanent plant cropping several years

Squash
See Pumpkin

Swede
'Acme'; small, sweet, yellow flesh; 'Marian', large, yeilow flesh, purple top, resistant clubroot and mildew

Sweetcorn
'Butter Imp', 75cm (30in), cobs 10-12cm (4-5in); 'Early Xtra Sweet', sweetest flavour; 'John Innes Hybrid', F1 hybrid, reliable, vigorous; 'Polar Vee', F1 hybrid, earliest; 'Sundance', 17cm (7in) long cobs, hardy; 'Tokay Sugar', F1

hybrid, white cobs, early

Swiss Chard
Silver or Seakale Beet; Rhubarb (Ruby) Chard, red stems

Tomato
Greenhouse: 'Eurocross BB', sets well in short days, leafmould resistant; 'Big Boy', a very large fruit, allow 3 trusses only; 'Golden Sunrise', a yellow; 'Grenadier', F1 hybrid, large, resistant leafmould and fusarium wilt; 'Ida', compact, early, resistant TMV, leafmould, fusarium and verticillium wilt; 'Tangella', orange, early; Greenhouse/outdoors: 'Marmande', large, good flavour (not g/house); 'Moneymaker', reliable setting, uniformly sized fruit; 'Outdoor Girl', early ripening; 'Sweet 100', cherry-size fruit, in long trusses, early; 'Tigerella', yellow striped on red, very good flavour; 'Tiny Tim', short plants, good for containers, fruit 2.5cm (1in) diameter; Outdoor bush: 'French Cross', F1 hybrid, heavy cropper; 'Pixie', 45cm (18in) tall, good in containers, small-medium fruit; 'Red Alert', 45-60cm (18-24in) tall, very early, good flavour, 2.5cm (1in) or more diameter fruit; 'Roma', plum shaped; 'Subarctic Plenty', withstands frost; 'The Amateur', reliable, heavy cropper, good flavour

Turnip
Early: 'Purple Top Milan', flat-globe, purple top; 'Milan White Forcing', round; maincrop: 'Green Globe', white, round, good for turnip-tops in spring; 'Golden Ball', round, keeps well; 'Model White', round; 'Veitch's Red Globe', round, white, red tops

Botanical and common names

Agaricus campestris	Mushroom	*Daucus carota*	Carrot
Allium ampeloprasum 'Porrum'	Leek	*Dioscorea* spp	Yam
		Eleocharis tuberosa	Water chestnut
Allium cepa 'Aggregatum'	Egyptian onion, tree onion	*Foeniculum vulgare* 'Dulce'	Florence fennel
A.c. 'Ascalonicum'	Shallot	*Helianthus tuberosus*	Jerusalem artichoke
A. fistoulosum	Welsh onion, Japanese bunching onion, ciboule	*Hibiscus esculentus*	Okra, ladies'-fingers, gumbo
Apium graveolens	Celeriac, knob celery, turnip rooted celery, German celery	*Ipomaea batatus*	Sweet potato
		Lactuca sativa	Lettuce
A. g. 'Dulce'	Celery	*Lotus tetragonolobus*	Asparagus pea
Asparagus officinalis	Asparagus	*Lycopersicon esculentum*	Tomato, love-apple, apple of Peru, Peruvian apple
Beta vulgaris 'Crassa'	Beetroot, beets		
B.v. 'Cicla'	Perpetual spinach/spinach beet/Swiss Chard/silver beet/ seakale beet	*Pastinacea sativa*	Parsnip
		Persea americana	Avocado pear
		Petroselinum crispum 'Tuberosum'	Hamburg parsley, turnip- rooted parsley
Brassica chinesis	Chinese cabbage, Chinese leaves	*Phaseolus coccineus*	Runner bean, pole bean, stick bean
B. napus 'Napobrassica'	Swede		
B. oleracea	Cabbage	*P. vulgaris*	French bean, kidney bean, haricot vert, shell bean, snap bean, string bean, haricot
B. o. 'Acephala'	Kale		
B. o. 'Botrytis', 'Cauliflora'	Cauliflower		
B. o. 'Bullata'	Savoy cabbage	*Pisum sativum*	Pea
B. o. 'Cymosa'	Sprouting broccoli	*Psalliota hortensis*	Mushroom
B. o. 'Gemmifera'	Brussels sprouts	*Raphanus sativus*	Radish
B. o. 'Gongyloides'	Kohlrabi	*Rheum rhaponticum*	Rhubarb
B. o. 'Italica'	Calabrese, broccoli	*Scorzonera hispanica*	Scorzonera, black salsify
B. rapa	Turnip	*Sechium edule*	Chayote, choko
Capsicum annum	Sweet pepper, capsicum	*Solanum melongena* 'Esculentum'	Aubergine, eggplant, brinjal, Jew's-apple
C. frutescens	Chilli	*S. tuberosum*	Potato, batata
Chichorium endivia	Endive, Batavian endive, escarole	*Spinacea oleracea*	Winter spinach
		S. o. 'Inermis'	Summer spinach
C. intybus	Chicory, Belgian endive	*Stachys affinis*	Chinese artichoke, crosnes
Colocasia antiquorum	Taro, eddoe, kandalla	*Tetragonia expansa*	New Zealand spinach
Cucumis sativus	Cucumber	*Tragopogon porrifolius*	Salsify, vegetable oyster
Cucurbita maxima	Winter squash	*Vicia faba*	Broad bean, fava bean, English bean
C. pepo	Pumpkin, marrow/marrow squash/vegetable squash		
Cynara scolymus	Globe artichoke	*Zea mays* 'Saccharata'	Sweetcorn, Indian corn, maize

Nutritional values

Notes to Nutritional Values Table

g = grams

mg = milligrams = $\frac{1}{1000th}$ of a gram

µ =

( ) = estimated value

— = no information available

Tr = trace, but of no dietetic significance

[1]Cabbage, carotene value is average figure, outer green leaves can contain 50 times as much

[2]Cabbage, white, vitamin C value – 20% of this is lost by shredding.

Vegetable	g Water	g Sugar	g Dietary Fibre	No. Calories	g Protein	g Fat	Minerals mg Na	K	Ca	N
ARTICHOKE, GLOBE boiled	84.4	—	—	15	1.1	Tr	15	330	44	
ARTICHOKE, JERUSALEM boiled	80.2	—	—	18	1.6	Tr	3	420	30	
ASPARAGUS boiled	92.4	1.1	1.5	18	3.4	Tr	2	240	26	
AUBERGINE raw, flesh only	93.4	2.9	2.5	14	0.7	Tr	3	240	10	
BEANS, BROAD boiled, no pods	83.7	0.6	4.2	48	4.1	0.6	20	230	21	2
BEANS, FRENCH boiled, pods and beans cut up	95.5	0.8	3.2	7	0.8	Tr	3	100	39	
BEANS, RUNNER pods trimmed, boiled	90.7	1.3	3.4	19	1.9	0.2	1	150	22	
BEETROOT boiled	82.7	9.9	2.5	44	1.8	Tr	64	350	30	
BRUSSELS SPROUTS boiled	91.5	1.6	2.9	18	2.8	Tr	2	240	25	
CABBAGE, RED raw	89.7	3.5	3.4	20	1.7	Tr	32	300	53	
CABBAGE, SAVOY boiled	95.7	1.1	2.5	9	1.3	Tr	8	120	53	
CABBAGE, SPRING boiled	96.6	0.8	2.2	7	1.1	Tr	12	110	30	
CABBAGE, WHITE raw	90.3	3.2	2.7	22	1.9	Tr	7	280	44	
CALABRESE	No figures available									
CARROT old, boiled	91.5	4.2	3.1	19	0.6	Tr	50	87	37	
CAULIFLOWER boiled, stalk/florets	94.5	0.8	1.8	9	1.6	Tr	4	180	18	
CELERIAC boiled	90.2	1.5	4.9	14	1.6	Tr	28	400	47	
CELERY raw	93.5	1.2	1.8	8	0.9	Tr	140	280	52	
CHICORY raw	96.2	—	—	9	0.8	Tr	7	180	18	
CUCUMBER raw, flesh only	96.4	1.8	0.4	10	0.6	0.1	13	140	23	
ENDIVE raw	93.7	1.0	2.2	11	1.8	Tr	10	380	44	
FENNEL	No figures available									
LEEK boiled, bulb only	90.8	4.6	3.9	24	1.8	Tr	6	280	61	

³Leek, carotene value is for bulb only; the leaves contain about 2000g.

⁴Lettuce, carotene value, as for cabbage.

⁵Potato, vitamin C value is percentage of raw potato C values; these vary between 30 and 8g/100g, from freshly-dug to stored 6-8 months.

⁶Rhubarb; with sugar, values are not appreciably altered except for:
sugar 11.4g
calories 45.

Carotene is given in the table as it is converted in the body into vitamin A and is the form in which it appears in vegetables; vitamin A itself is not present in them.

B1 = thiamin; deficiency results in beriberi.

B2 = riboflavin.

B6 = pyrodoxine and its forms.

Folic acid is also part of the vitamin B complex; figures for it are not given here as it is likely that the figures available will have to be considerably revised, and cannot be regarded as valid. However, raw, green, leafy vegetables contain appreciable quantities, which are readily lost in boiling.

Fe	Cu	Zn	S	Cl	per 100g µg Carotene	µg D	Vitamins g B1	B2	Nicotinic acid	B6	µg B12	C
0.5	0.09	—	16	84	90	0	0.07	0.03	0.9	0.07	0	30
0.4	0.12	0.1	22	68	(Tr)	0	0.10	Tr	—	—	0	2
0.9	0.20	0.3	47	31	500	0	0.10	0.08	0.8	0.04	0	20
0.4	0.08	—	9	61	Tr	0	0.05	0.03	0.8	0.08	0	5
1.0	0.43	—	27	14	250	0	(0.10)	0.04	3.0	—	0	15
0.6	0.10	0.3	8	11	400	0	0.04	0.03	0.3	0.06	0	5
0.7	0.05	0.3	—	5	400	0	0.03	0.07	0.5	0.04	0	5
0.4	0.08	0.4	22	76	Tr	0	0.02	0.04	0.1	0.03	0	5
0.5	0.05	0.4	78	16	400	0	0.06	(0.10)	0.4	0.17	0	—
0.6	0.09	0.3	68	45	(20)	0	0.06	0.05	0.3	0.21	0	55
0.7	0.07	0.2	30	9	300	0	0.03	0.03	0.2	0.10	0	15
0.5	0.07	0.2	27	6	500	0	0.03	0.03	0.2	0.10	0	25
0.4	(0.03)	0.3	—	23	(Tr)	0	0.06	0.05	0.3	0.16	0	40²
0.4	0.08	0.3	5	31	¹2000	0	0.05	0.04	0.4	0.09	0	4
0.4	(0.03)	0.2	—	14	30	0	0.06	0.06	0.4	0.12	0	20
0.8	0.13	—	13	23	0	0	0.04	0.04	0.5	0.10	0	4
0.6	0.11	0.1	15	180	Tr	0	0.03	0.03	0.3	0.10	0	7
0.7	0.14	0.2	13	25	Tr	0	0.05	0.05	0.5	0.05	0	4
0.3	0.09	0.1	11	25	Tr	0	0.04	0.04	0.2	0.04	0	8
2.8	0.09	—	26	71	2000	0	0.06	0.10	0.4	—	0	12
2.0	0.09	(0.1)	49	43	40³	0	0.07	0.03	0.4	0.15	0	15

Vegetable	g Water	g Sugar	g Dietary Fibre	No. Calories	g Protein	g Fat	Na	K	Ca	M
							Minerals mg			
LETTUCE raw	95.9	1.2	1.5	12	1.0	0.4	9	240	23	
MARROW & COURGETTE boiled	97.8	1.3	0.6	7	0.4	Tr	1	84	14	
MUSHROOM raw, flesh and stem	91.5	0	2.5	13	1.8	0.6	9	470	3	1
OKRA raw (literature sources)	90.0	2.3	(3.2)	17	2.0	Tr	·7	190	70	6
ONION raw	92.8	5.2	1.3	23	0.9	Tr	10	140	31	
PARSNIP boiled	83.2	2.7	2.5	56	1.3	Tr	4	290	36	1
PEA boiled	80	1.8	5.2	52	5.0	0.4	Tr	170	13	2
PEPPER sweet, raw	93.5	2.2	Tr	15	0.9	0.4	2	210	9	1
POTATO old, boiled, peeled	80.5	0.4	1.0	80	1.4	0.1	3	330	4	1
PUMPKIN raw, flesh only	94.7	2.7	0.5	15	0.6	Tr	1	310	39	
RADISH raw	93.3	2.8	1.0	15	1.0	Tr	59	240	44	1
RHUBARB stewed, no sugar	94.6	0.9	2.4	6	0.6	Tr	2	400	93	1
SPINACH boiled	85.1	1.2	6.3	30	5.1	0.5	120	490	600	5
SPROUTING BROCCOLI	No figures available									
SQUASH	See pumpkin, figures likely to be similar									
SWEDE boiled	91.6	3.7	2.8	18	0.9	Tr	14	100	42	
SWEETCORN boiled, kernels	65.1	1.7	4.7	123	4.1	2.3	1	280	4	4
SWISS CHARD	No figures available									
TOMATO raw	93.4	2.8	1.5	14	0.9	Tr	3	290	13	1
TURNIP boiled	94.5	2.3	2.2	14	0.7	0.3	28	160	55	

Other foods (for comparison)

Vegetable	g Water	g Sugar	g Dietary Fibre	No. Calories	g Protein	g Fat	Na	K	Ca	M
BREAD wholemeal	40	2.1	8.5	216	8.8	2.7	540	220	23	9
CHEESE cheddar type	37	Tr*		406	26.0	33.5	610	120	800	2
CHOCOLATE plain	0.6	56.5	—	525	4.7	29.2	11	300	38	10
EGG boiled	74.8			147	12.3	10.9	140	140	52	
FISH cod, steamed	79.2			83	18.6	0.9	100	360	15	2
MEAT beef, topside, roast, lean and fat, 88% lean	60.2			214	26.6	12.0	48	350	6	2
MILK cow's, fresh, whole	87.6	4.7*		65	3.3	3.8	50	150	120	1

N.B. Figures for milk from Milk Marketing Board and literature sources; the small losses due to pasteurisation make no difference to the average composition for milk, except for vitamin C, of which 25% is lost as a result, but see also note c below.

* = lactose
b = milk not exposed to light; loss on exposure to sunlight of 10% per hour.

c = as delivered to home; after 12 hours is 1.0mg and 0.5mg after 24 hours.
d = summer

P	Fe	Cu	Zn	S	Cl	per 100g μg Carotene	μg D	Vitamins g B1	B2	Nicotinic acid	B6	μg B12	C
7	0.9	(0.03)	0.2	—	53	1000[1]	0	0.07	0.08	0.3	0.07	0	15
3	0.2	0.03	0.2	6	14	30	0	Tr	(Tr)	0.2	0.03	0	2
0	1.0	0.64	0.1	34	85	0	0	0.10	0.40	4.0	0.10	0	3
0	1.0	0.19	—	30	41	90	0	0.10	0.10	1.0	0.08	0	25
0	0.3	0.08	0.1	51	20	0	0	0.03	0.05	0.2	0.10	0	10
2	0.5	0.10	0.1	15	33	Tr	0	0.07	0.06	0.7	0.06	0	10
3	1.2	0.15	0.5	44	8	300	0	0.25	0.11	1.5	0.10	0	15
6	0.4	0.07	0.2	—	18	200	0	Tr	0.03	0.7	0.17	0	100
9	0.3	0.11	0.2	22	41	Tr	0	0.08	0.03	0.8	0.18	0	50[b]
9	0.4	0.08	(0.2)	10	37	1500	0	0.04	0.04	0.4	0.06	0	5
7	1.9	0.13	0.1	38	19	Tr	0	0.04	0.02	0.2	0.10	0	25
9	0.4	0.12	—	7	81	55	0	Tr	0.03	0.3	0.02	0	8
3	4.0	0.26	0.4	86	56	6000	0	0.07	0.15	0.4	0.18	0	25
3	0.3	0.04	—	31	9	Tr	0	0.04	0.03	0.8	0.12	0	17
0	0.9	0.15	1.0	—	14	(240)	0	0.20	0.08	1.7	0.16	0	9
	0.4	0.10	0.2	11	51	600	0	0.06	0.04	0.7	0.11	0	20
	0.4	0.04	—	21	31	0	0	0.03	0.04	0.4	0.06	0	17
0	2.5	0.27	2.0	81	860	0	0	0.26	0.08	3.9	0.14	0	0
0	0.4	0.03	4.0	230	1060	205	0.26	0.04	0.50	0.10	0.080	1.5	0
0	2.4	0.70	0.2	—	100	(40)	0	0.07	0.08	0.4	(0.02)	0	0
0	2.0	0.10	1.5	180	160	Tr	1.75	0.08	0.45	0.07	0.10	1.7	0
0	0.5	0.10	0.5	210	120	Tr	Tr	(0.09)	(0.09)	(2.1)	(0.37)	(3)	Tr
6	2.6	0.13	4.9	—	51	Tr	Tr	0.07	0.31	5.7	0.29	2	0
5	0.05	0.02	0.35	30	95	22[d]	0.03[d]	0.04	0.19[b]	0.08	0.04	0.3	1.5[c]

Vegetable	Sowing date	Planting date	Time to germinate (days)	Prick out	Thin	Transpl
ARTICHOKE, GLOBE		mid-spring, mid-autumn				
ARTICHOKE, JERUSALEM		late winter–mid-spring				
ASPARAGUS		mid–late spring				
AUBERGINE	late winter–early spring		10–20	yes		
BEAN, BROAD	spring, late autumn		10–20			
BEAN, FRENCH	early spring–midsummer		10–14			
BEAN, RUNNER	mid–late spring		10–15			
BEETROOT	mid-spring–early summer		15–24		yes	
BRUSSELS SPROUT	early–mid-spring		7–12		yes	yes
CABBAGE	mid–late spring, mid–late summer		7–12		yes	yes
CABBAGE, CHINESE	mid–late summer		7–12		yes	
CALABRESE	early–late spring		7–12		yes	yes
CARROT	early spring–midsummer		14–24		yes	
CAULIFLOWER	early–late spring		7–12		yes	yes
CELERIAC	mid-spring		10–20		yes	yes
CELERY	mid-spring		18–28		yes	yes
CHICORY	early–late summer		6–12		yes	
CUCUMBER	mid–late spring		6 (average)			yes
ENDIVE	mid-spring–early autumn		5–21		yes	
FENNEL, FLORENCE	midsummer		6–12		yes	
LEEK	early–mid-spring, early summer		9–21		yes	yes
LETTUCE	early spring–mid-autumn		10–18		yes	
MARROW	mid–late spring		6–12			yes
MUSHROOM	best spring–midsummer		5 weeks			
OKRA	spring		6–9			yes
ONION	early–mid-spring, late summer	early to mid-spring (sets)	21		yes	
PARSNIP	early–mid-spring		21–28		yes	
PEA	early spring–midsummer, mid to late autumn		7–20			
PEPPER	late winter–mid-spring		14–20	yes	yes	yes
POTATO		early–mid-spring				
PUMPKIN	mid–late spring		5			yes
RADISH, SUMMER	early spring–late summer		4–10		yes	

Pot	Protect till final planting	Sowing/planting depth cm/in	Spacing		Remarks
			In row cm/in	**Between rows cm/in**	
	no	10 (4)	120 (48)	90–120 (36–48)	crown just above soil level
	no	10–15 (4–6)	60 (24)	90 (36)	sprouting takes two–four weeks
	no	23 (9)	45 (18)	90 (36)	if cold and wet, plant late spring
yes	yes	0.6 (¼)	60 (24)	60 (24)	also grow in 23cm (9in) final pots
	no	5 (2)	20–25 (8–10)	38–45 (15–18)	sow in autumn in mild areas, protect in winter
	until late spring	5 (2)	10 (4)	45 (18)	
	until late spring	5 (2)	23 (9)	45 (18)	
	no	2.5 (1)	7–15 (3–6)	20 (8)	station-sow 2.5cm (1in) apart
	no	1.5 (½)	45 (18)	45–75 (18–30)	
	no	1.5 (½)	30–45 (12–18)	30–45 (12–18)	space spring cabbage 10 × 30cm (4 × 12in); choose right vars. for season
	autumn	1.5 (½)	30 (12)	30 (12)	station sow at 10cm (4in)
	early–mid-spring	1.5 (½)	30 (12)	30 (12)	
	no	1.5 (½)	5–15 (2–6)	15–30 (6–12)	closer spacing for earlies
	no	1.5 (½)	60–75 (24–30)	60–75 (24–30)	space small summer cauliflowers 15cm (6in) apart each way; choose right vars. for season
	yes	0.6 (¼)	30 (12)	45 (18)	plant without root disturbance
	yes	0.6 (¼)	23–30 (9–12)	38 (15)	self-blanching 23 × 23cm (9 × 9in)
	no	1.5 (½)	15–20 (6–8)	30 (12)	
yes	yes	1.5 (½)	75 (30)	75 (30)	indoor vars. spaced 45cm (18in) apart
	no	1.5 (½)	23–30 (9–12)	30 (12)	use close spacing for curly vars.
	no	0.6 (¼)	20 (8)	30 (12)	thin as early as possible
	no	0.6 (¼)	10–15 (4–6)	30 (12)	trim back roots/leaf tips
	early-mid-spring	0.3–0.6 (⅛–¼)	15–30 (6–12)	20–30 (8–12)	choose right vars. for season
yes	yes	2.5 (1)	75 (30)	75 (30)	trailing vars., 120 × 60cm (48 × 24in)
	all year	2.5 (1)			can be spawned all year with protection
yes	yes	1.5 (½)	45 (18)	60 (24)	keep free of frost
	no	0.6 (¼)	7–20 (3–8)	23 (9)	adjust plant spacing according to size of bulb wanted
	early spring	1.5 (½)	15 (6)	20–30 (8–12)	station sow
	early–mid-spring	5 (2)	7 (3)	60–150 (24–60)	staggered double rows 10cm (4in) apart
yes	yes	0.6 (¼)	45 (18)	45 (18)	23cm (9in) diameter final pot size if pot grown
	yes	10–15 (4–6)	30–38 (12–15)	60–75 (24–30)	put to sprout mid–late winter; use close spacing for earlies
yes	yes	2.5 (1)	90–120 (36–48)	90–120 (36–48)	
	no	1.5 (½)	7–23 (3–9)	15–30 (6–12)	thin as soon as possible after germination; sow thinly; use larger spacings for winter vars.

Vegetable	Sowing date	Planting date	Time to germinate (days)	Prick out	Thin	Transpl
RHUBARB		late autumn, early spring				
SPINACH, WINTER, SUMMER & PERPETUAL	spring, late, summer		10–20		yes	
SPINACH, NEW ZEALAND	late spring		10–20			
SPROUTING BROCCOLI	mid–late spring		7–12		yes	yes
SQUASH	mid–late spring		5			yes
SWEDE	early summer		6–10		yes	
SWEETCORN	mid-spring		10–15		yes	yes
SWISS CHARD	late spring–early summer		6–14		yes	
TOMATO, INDOOR	late winter–mid-spring		8–11	yes		yes
TOMATO, OUTDOOR	mid-spring		8–11	yes		yes
TURNIP	mid–late summer		6–14		yes	

Harvesting

● = fresh
— = stored

	Jan	Feb	Mar	Apr	May	Jun	Jul	Aug	Sep	Oct	Nov	Dec	Crop yields (average)
ARTICHOKE, GLOBE						●	●	●	●				5 large heads, 10 small/plant
ARTICHOKE, JERUSALEM	●	●	●								●	●	1–1½kg (2–3lb)/plant
ASPARAGUS				●	●	●							15–20 spears/plant
AUBERGINE								●	●	●			4 fruits/plant (purple vars.)
BEAN, BROAD						●	●	●					2.5kg (5lb)/3-metre (10ft) row
BEAN, FRENCH							●	●	●				3–4kg (6–8lb)/3-metre (10ft) row
BEAN, RUNNER							●	●	●				0.75–1kg (1½–2lb)/plant
BEETROOT	—	—	—		●	●	●	●	●	—	—	—	125–200g (4–7oz)/root, depending on type
BRUSSELS SPROUTS	●	●	●						●	●	●	●	¾–1kg (1½–2lb)/plant
CABBAGE	●	●	●	●	●	●	●	●	●	●	●	●	½–2kg (1–4lb)/head, depending on var.
CABBAGE, CHINESE									●	●	●	●	1–2kg (2–4lb)/head
CALABRESE								●	●	●			½kg (1lb)/plant
CARROT	—	—	—		●	●	●	●	●	●	—	—	3–4kg (6½–8½lb)/3-metre (10ft) row
CAULIFLOWER		●	●	●	●	●	●	●	●	●	●		½–1¼kg (1–2lb)/plant, of curd
CELERIAC	—	—							●	●	●		250g (½lb)/plant
CELERY	●						●	●	●				¾–1kg (1½–2lb)/plant
CHICORY	●	●	●	●					●	●	●	●	90–150g (3–5oz)/head
CUCUMBER							●	●	●	●			10–15/plant, ridge; 25/plant indoor
ENDIVE	●	●	●					●	●	●	●	●	9–13 heads/3-metres (10ft) row
FENNEL, FLORENCE								●	●	●			each bulb 125–250g (¼–½lb)
LEEK	●	●	●	●						●	●	●	weight trimmed leek, 125–250g (¼–½lb) each
LETTUCE	●	●	●	●	●	●	●	●	●	●	●	●	10–15 heads/3-metre (10ft) row
MARROW & COURGETTE						●	●	●	●	—	—		marrow ½–2kg (1–4lb); courgette 56–84g (2–3oz); 5 plants sufficient for family

Pot	Protect till final planting	Sowing/planting depth cm/in	In row cm/in	Between rows cm/in	Remarks
	no	about 20 (8)	90 (36)	90 (36)	cover crown with 5cm (2in) soil
	no	2.5 (1)	23–30 (9–12)	30–38 (12–15)	winter and 2nd sowing of perpetual spinach in late summer; perpetual spinach at wider spacings
	no	2 (¾)	30 (12)	60 (24)	four or five plants sufficient for family
	no	1.5 (½)	60 (24)	60 (24)	
yes	yes	2.5 (1)	90–120 (36–48)	90–120 (36–48)	
	no	1.5 (½)	23 (9)	38 (15)	sow thinly; sow late spring in cold areas
	yes	2 (¾)	30–45 (12–18)	30–45 (12–18)	plant in block; use wider spacing for tall vars.
	no	2.5 (1)	30 (12)	38 (15)	station sow 10cm (4in) apart
yes	yes	1.5 (½)	45 (18)	60 (24)	plant out late spring
yes	yes	1.5 (½)	45 (18)	75 (30)	
	no	1.5 (½)	12–23 (5–9)	23–30 (9–12)	thin very early; sow earlies early spring–early summer at narrow spacing

● = fresh
— = stored

	Jan	Feb	Mar	Apr	May	Jun	Jul	Aug	Sep	Oct	Nov	Dec	Crop yields (average)
MUSHROOM	●	●	●	●	●	●	●	●	●	●	●	●	250–1000g (½–2lb)/sq ft
OKRA							●	●	●				10–20 pods/plant
ONION	●	●	●	●	●	●	●	●	●	●	●	●	4kg (9lb)/3-metre (10ft) row (bulb onions)
PARSNIP	● —	● —	● —							●	●	● —	180–500g (6–16oz)/root
PEA						●	●	●	●				250g (8oz)/plant
PEPPER, SWEET							●	●	●				4–8 fruits/plant
POTATO	—	—	—	—		●	●	●	●		—	—	early: 750g (1½lb)/plant, maincrop: 1250g (2½lb)/plant
PUMPKIN									●	●	● —		4 fruits/plants, each 7kg (15½lb)
RADISH	● —	●		●	●	●	●	●	●	●	● —	● —	summer vars. 2kg (4½lb)/3-metre (10ft) row; winter vars. 5kg (11lb)/3-metre (10ft) row
RHUBARB				●	●	●	●						3kg (6lb)/plant
SPINACH	●	●	●	●	●	●	●	●	●	●	●	●	summer: 140g (5oz)/plant; winter/perpetual: 224g (8oz)/plant; New Zealand 336g (12oz)/plant
SPROUTING BROCCOLI			●	●	●								½kg (1lb)/plant
SQUASH		—							●	●	—	—	varies greatly depending on var.; 6 1kg (2lb) fruits/plant average
SWEDE	● —	● —	● —							●	●	● —	1kg (2½lb)/root
SWEETCORN									●	●			2 cobs/plant
SWISS CHARD			●	●			●	●	●	●	●	●	330g (¾lb)/plant
TOMATO							●	●	●	●	●		4kg (9lb)/plant indoors; 2¼kg (5lb)/plant outdoors
TURNIP	● —	● —				●	●	●	●	●	● —	● —	early: 120–150g (4–5oz) each root; maincrop: 500g (1lb) each

Storing and preserving vegetables

There are various methods of keeping vegetables for out-of-season use; the simplest concerns the root crops, the majority of which can be stored by leaving them in the ground until needed, taking care to put a mulch of bracken or straw over them to prevent the ground freezing and becoming impossible to dig. Otherwise dig them up, clean and trim, and place in layers in very slightly moist sand or peat in boxes proof against mice, and keep in a frostproof dark place. An old method, especially for potatoes, is to make a clamp, in which the roots or tubers are placed in layers in a sloping mound covered with straw, and then with a layer of packed-down soil, smoothed on the outside for rain drainage, and with an outlet of straw at the top for ventilation.

In ground
Jerusalem artichoke, celeriac, kohlrabi, parsnip, radish, salsify, scorzonera, swede, turnip (can also be in boxes but better flavoured if left in ground).
In boxes/clamps
Beetroot, carrot, celeriac (in cold gardens), potato.
Hanging in frostproof shed
Cabbage, cauliflower, marrow, onion, pumpkin, squash (first two upside down with roots on).

DRYING VEGETABLES
Although most vegetables are better preserved by bottling or freezing, onions, peas and beans can be dried. Overgrown beans, peas and haricots should be left on the vine until the pods are yellow. The seeds are then removed and slowly dried in a cool dry place. Plaits of onions can be hung indoors in a cool, dry, airy place but it is also useful to have a supply of dried onion rings in the kitchen. These should be blanched in boiling water for about 30 seconds, patted dry with kitchen towels, then dried in single layers in a cool oven (maximum temperature 65°C/150°F/gas mark ¼) until crisp and dry. It will take about 3 hours. They can be added straight to casseroles and stews. Mushrooms can also be dried by removing the stalks, threading them on string about 5cm (2in) apart and hanging them to dry in a warm, airy place where they will take about 2 days to dry.

SALTING
Beans – and cabbage in the form of sauerkraut – are really the only vegetables suitable for salting. Coarse cooking salt or sea salt should be used (not refined table salt) in the proportion of 500g/1lb salt to 1½kg/3lb beans. Pack the beans and salt in alternate layers in unglazed earthenware or glass jars, starting and ending with salt. Do not let the salt come into contact with metal. Compress the mixture with an inverted plate until the beans shrink and the salt dissolves into a brine. Then cover the jar tightly. Take out quantities as required, wash under cold running water and leave to soak in cold water for about 2 hours to draw out the salt. Give them a final rinse before boiling as usual in unsalted water.

BOTTLING
Among the vegetables suitable for bottling are asparagus, beans, carrots, cauliflower, celery, button mushrooms, peas, tomatoes and vegetable macedoine. Bottling needs special care and equipment so it is worthwhile buying a book which gives detailed instructions about the process. Green vegetables, such as peas and beans, lose a good deal of their colour when bottled and can look old and faded, although their flavour is unimpaired. However, asparagus, cauliflower, carrots and tomatoes all bottle well.

PICKLING AND CHUTNEYS
The great number of vegetables which can be pickled rely on a spiced vinegar marinade to preserve them. Some are pickled raw, such as onion, cabbage, cauliflower and cucumber, others after cooking, such as beetroot, carrots, marrow and tomato. While pickling basically aims at preserving the colour, shape and texture of the original vegetable, chutneys are composed of chopped vegetables blended with spices and other ingredients. Pumpkin, tomatoes, pepper, marrow, beetroot and rhubarb make excellent chutneys. Sauces, ketchups (notably mushroom and tomato) and relishes can also be easily made from many vegetables. Home preserving of this kind is not difficult and, apart from freezing, is probably the most satisfactory and satisfying way of using up a harvest glut. A reliable recipe book of pickles and chutneys will not only give instructions but also an imaginative range of ideas for using vegetables in these ways.

FREEZING
Nearly all vegetables can be frozen without too much trouble, the only exception being salad vegetables whose high water content means that they go soggy. However some vegetables such as Jerusalem artichokes, aubergine and marrow or courgette are best cooked and frozen as part of a prepared dish. Jerusalem artichokes should be made into soup or puréed, then frozen; aubergines and courgettes can be made into ratatouille and frozen. All vegetables for freezing should be prepared as for cooking and then blanched. Blanching times are given opposite. After blanching, the vegetables should be plunged into cold water, then drained thoroughly and dried on paper towels before freezing. Properly frozen, vegetables should keep in good condition for up to 9 months or more.

*** freezes very well
** will freeze
* freeze as cooked dish

Vegetable	Suitability For Freezing	Preparation and Freezing Instructions	Blanching time
ARTICHOKE, GLOBE	**	Remove outer leaves and choke, blanch in acidulated water, cool; pack in rigid container	5-7 min
ARTICHOKE, JERUSALEM	*	Cook as soup or purée before freezing in bags or rigid container	
ASPARAGUS	**	Trim to even lengths, blanch, cool, tie in bundles; layer in rigid container separating layers with waxed paper	thin stems 2 min thick stems 4 min
AUBERGINES	*	Cook in cubes with onions or tomatoes as part of ratatouille before freezing in bags or rigid containers	
BEANS, BROAD	***	Blanch, cool and dry pods; pack in freezer bags	2 min
FRENCH	***	Top and tail, blanch, dry; pack in freezer bags	2 min
RUNNER	***	Top and tail, string cut into 2.5cm (1 inch pieces); blanch, dry; pack in freezer bags	2 min
BEETROOT	*	Cook as soup or in cubes with apples before freezing	
BRUSSELS SPROUTS	***	Remove outer leaves; divide into groups of uniform size for blanching separately; pack in freezer bags	3-4 min according to size
CABBAGE	*	Braise red cabbage before freezing in bag or rigid container	
CALABRESE	***	Break into sprigs. Blanch, cool; pack in rigid container separating layers with waxed paper	3min
CARROTS	***	Trim and peel, blanch, cool; pack whole or sliced in freezer bags	3 min
CAULIFLOWER	***	Break into sprigs, blanch in acidulated water, cool; pack in rigid container in layers separated by wax paper	3 min
CELERIAC	**	Trim, peel, cut into cubes, slices or sticks; blanch, cool; pack in rigid container separating layers with wax paper, or bags	4 min
CELERY STALKS	**	Trim, cut into 2.5cm (1 inch) pieces, blanch, cool; pack into freezer bags	3-4 min
HEARTS	**	Blanch, cool; pack into freezer bags	2 min
COURGETTE (see Marrow)			
CUCUMBER	*	Cook as braised, or in soup before freezing	
FENNEL, FLORENCE	**	Trim, cut into 2.5cm (1 inch) pieces, blanch; cool, pack into freezer bags	3 min
LEEKS	***	Remove outer leaves, wash, cut into 2.5cm (1 inch) pieces; blanch, cool; pack into freezer bags	2 min
MARROW	*	Cook as braised or in stew before freezing	
MUSHROOM, BUTTON	**	Wipe clean, blanch, cool; pack into rigid container	4 min
PARSNIP	***	Trim, peel, dice or slice, blanch; pack into freezer bags	2 min
PEAS, GARDEN	***	Shell, blanch; pack in freezer bags	2 min
MANGETOUT	***	String, blanch; pack in rigid container	3 min
PEPPERS, SWEET	**	Remove stalk, de-seed, cut into pieces or strips, blanch; pack in rigid container	2-4 min
POTATOES	**	Peel or scrape, cut into dice or chips, blanch; pack into freezer bags	3-4 min
SPINACH	***	Remove leaves from stalks, wash, blanch, squeeze out excess moisture; pack in freezer bags or rigid container	2 min
SPROUTING BROCCOLI	***	Trim off leaves, divide into sprigs, blanch, cool; pack into rigid containers	2 min
SWEETCORN ON THE COB	***	Remove husk and silks, trim, blanch, wrap each in foil; pack in freezer bags	4-6 min
KERNELS	***	Scrape kernels from blanched cobs; pack in freezer bags	
TOMATOES	***	Freeze as purée or juice; pack in rigid containers	
TURNIPS	**	Trim, peel and dice, blanch; pack in freezer bags	2 min

Suppliers and useful addresses

General suppliers

John Barber (Hertford) Ltd.,
Seedsmen,
2 St Andrew St., Old Cross,
Hertford SG14 1JD

J. W. Boyce, Seedsmen,
67 Station Rd., Soham, Ely,
Cambs. CB7 5ED

Dickson Brown & Tait,
Attenburys Lane, Timperley,
Altrincham, Cheshire
WA14 5QL

Samuel Dobie & Son Ltd.,
Upper Dee Mills, Llangollen,
Clwyd LL20 8SD

S. E. Marshall & Co. Ltd.,
Regal Rd., Wisbech, Cambs.
PE13 2RF

Suttons Seeds Ltd.,
Hele Rd., Torquay, Devon
TQ2 7QJ

Thompson & Morgan,
London Rd., Ipswich, Suffolk
IP2 0BA

Unwins Seeds Ltd.,
Histon, Cambridge CB4 4LE

Societies and association

National Vegetable Society
c/o W. R. Hargreaves, 29
Revidge Road, Blackburn,
Lancs. BB2 6JB

Royal Horticultural Society,
Vincent Square, London
SW1P 2PE

Northern Horticultural Society,
c/o R. E. Shersby, Harlow Car
Gardens, Harrogate HG3 1QB

Henry Doubleday Research
Association,
Ryton-on-Dunsmore, Coventry
CV8 3LG

National Vegetable Research
Station,
Wellesbourne, Warwick
CV35 9EF

The Royal Horticultural Society
has a model vegetable garden
at its garden at Wisley in Surrey.

Specialist suppliers

Michael Bennett,
Long Compton, Shipston-on-
stour, Warks. CW36 5JN
(asparagus, globe artichokes)

John Chambers,
15 Westleigh Rd., Barton
Seagrave, Kettering, Northants
NN15 5AJ (wild vegetables)

Chase Organics,
Seed Division, Gibraltar House,
Govett Ave., Shepperton,
Middx. TW17 8AQ

Chiltern Seeds,
Bortree Stile, Ulverston,
Cumbria LA12 7PB (wide range
of unusual vegetable seeds)

Garlic Farm,
58 Churchill Rd., Brislington,
Avon BS4 3RW (garlic, Italian
and Greek vegetables,
Jerusalem artichokes, outdoor
tomatoes)

Henry Doubleday Research
Association,
Ryton-on-Dunsmore, Coventry
CV8 3LG (organically grown
seeds and plants)

Regal Lodge,
Kentford, Newmarket, Suffolk
CB8 7QB (asparagus, globe
artichokes)

W. Robinson & Sons Ltd.
Sunny Bank, Forton, Preston,
Lancs. PR3 0BN (onions, celery,
tomatoes)

Scotts Nurseries (Merriott) Ltd.,
Merriott, Somerset TA16 5PL
(unusual vegetables)

Suffolk Herbs,
Sawyers Farm, Little Cornard,
Sudbury, Suffolk CO10 0PF
(wide range of unusual
vegetable seeds)

Specialist potato-seed suppliers (tubers)

Arbroath Flower Centre,
31, Fisheracre, Arbroath,
Tayside, Scotland DD11 1LE

Glossary of gardening terms

Acid soil: see soil testing.

Alkaline soil: see soil testing.

Annual: Plant that completes a cycle of development from the germination of the seed through growth and death in a single season. Lettuce and radishes are annuals.

Biennial: Plant that requires two seasons to complete the growth cycle from germination of seed through maturity and death. Cabbages and other brassicas are biennials.

Blanching: Exclusion of light from the stems of some vegetables during their later stages of growth so that they remain pale in colour and more delicately flavoured. Chicory is frequently blanched.

Bolting: Premature running to seed which affects lettuce in particular. Many non-bolting varieties of vegetables are now available.

Brassicas: Members of the cabbage family, including Brussels sprouts, broccoli, cauliflower, kale and kohlrabi and close relations like swedes, turnips and radishes. Brassicas need special care and attention since they are attacked by a wide range of pests and diseases.

Cloche: Glass or plastic covering which forces a plant into early growth.

Compost: Vegetable matter left until well-rotted, then added to the soil as plant food.

Crop rotation: A growing system devised to make sure that no member of a plant family is grown in the same soil more than once in 3 years, preferably 4. This gives plant diseases a chance to die out.

Cultivar: A cultivated variety of a plant, different in some respects from the naturally occurring botanical variety.

Digging: Double digging is a method of cultivation to a soil depth of 2 spades instead of the normal one spade depth. It is most worthwhile for growing deeply rooted crops such as parsnips or runner beans.

Drill: A shallow furrow made in the soil, usually up to 5cm (2in) deep, in which seeds are sown.

Earthing up: Drawing soil around the stems of plants to provide protection or to exclude light from stems or tubers.

Foliar feeding: Method of nourishing a plant by applying a fertilising solution on to the plant's leaves.

Forcing: Excluding light or providing warm conditions in order to encourage premature growth of a plant.

Hardening off: The gradual introduction of outdoor conditions to tender plants nurtured indoors in warmth.

Intercropping: Growing quick-maturing plants between other vegetables that grow more slowly. Lettuce, for example, is frequently intercropped with slow-maturing plants.

Legume: Vegetable that produces pods e.g. peas and beans.

Mulch: Layer of material on the soil around the plant used to conserve water, prevent weed growth and provide a cool root run. If rotted organic matter is used, it will supply plant nutrient gradually and maintain a good soil structure.

Overwintering: Protecting plants from autumn through to spring to help them survive cold weather, usually by putting them in warmth or under cloches.

Prick out: To plant out seedlings from their germinating container into a larger one in which they are spaced out, usually about 5cm (2in) each way, to allow unrestricted growth.

Set: Seed bulbs of onions and shallots used for planting are called sets.

Station sowing: Sowing in specially prepared holes rather than drills.

Spit: depth of a garden spade or fork, as for normal digging.

Thin: To remove excess seedlings to make room for the remaining plants to mature.

Tilth: Fine, crumbly texture of soil, necessary when sowing seeds or small plants.

Transplanting: Moving plants raised in a seed bed or under glass to their permanent growing position.

Soil testing (for pH level): Measuring the degree of acidity or alkalinity in the soil, with the aid of a soil testing kit.

Viable: Alive or able to live. Of plants, a seed which still has the capacity to germinate.

BOOKS FOR FURTHER READING

Be your own vegetable doctor
D. G. Hessayon
Pan Britannica Industries Ltd
1978

A Calendar of Gardeners Lore (historical)
Susan Campbell
Century Publishing Co. Ltd 1983

Companion Planting (organic)
Gertrud Franck
Thorsons Publishers Ltd 1983

Edible Ornamental Garden, The
John E. Bryan & Coralie Castle
Pitman Ltd 1976

Food Scandal, The
Caroline Walker & Geoffrey Cannon
Century Publishing Co Ltd 1984

Good Things (cookery, mainly fruit and vegetables)
Jane Grigson Ltd
Penguin Books 1977

Grow it and Cook it
Denis Wood & Kate Crosby
Faber & Faber Ltd 1975

Incredible Heap, The (compost heap making)
Chris Catton & James Gray
Pelham Books Ltd 1983

Kitchen Garden, The (historical)
David C. Stuart
Robert Hale Ltd 1984

Know and Grow Vegetables
J. K. A. Bleasdale, P. J. Salter & others
Oxford University Press 2 vols
Vol 1 1979   Vol 2 1982

Manual of Nutrition
Ministry of Agriculture Fisheries & Food
HM Stationery Office 1985

Medieval English Gardens (historical)
Teresa McLean
Collins 1981

Organic Gardening
Lawrence D. Hills
Penguin 1977

Vegetable Pest & Disease Control the Organic Way
Lawrence D. Hills
Henry Doubleday Research Assocn 1985

Pests, Diseases & Disorders of Garden Plants, Collins Guide to
Stefan Buczacki, Keith Harris
Collins 1981

The Salad Garden
Joy Larkcom
Windward 1984

The Vegetable Garden
Vilmorin-Andrieux
English edition John Murray 1885
Facsimile reprint 1976

Vegetable Garden Displayed, The
Royal Horticultural Society
Rev. 1981

Vegetables Naturally (organic)
Jim Hay
Century Publishing Co Ltd 1985

Vegetarian Kitchen
Sarah Brown
British Broadcasting Corporation 1985

Index

ACKNOWLEDGEMENTS
The publishers would like to thank the following individuals and organizations for their help in producing this book:
The Controller of Her Majesty's Stationery Office for permission to reproduce nutritional values chart; Henry Doubleday Research Association; Suttons Seeds; Thompson & Morgan; Unwin Seeds.

Picture credits
A-Z Botanical 132 (bottom), 134 (t.left), 135 (t.left), 138, 139 (t.right & middle), 140 (t.right)
Ann Bonar 25, 39, 82 (inset), 94 (inset), 143 (right)
Bruce Coleman 48 (inset), 74 (inset), 92 (inset), 133, 140 (bottom)
Tom Deas 1, 2/3, 20, 24 (middle), 29 (below), 42 (top), 43 (left), 54 (inset), 56 (inset), 60 (inset), 64 (inset), 66 (inset), 68 (inset), 70 (inset), 72 (inset), 78 (inset), 80 (inset), 84 (inset), 98 (inset), 108 (inset), 112 (inset), 114 (inset), 118 (inset), 126 (inset), 140 (middle left), 142 (left)
Charles Nicholas 24 (top), 47 (inset), 144
Harry Smith Collection 134 (t.left), 139 (t.left)
Sutton Seeds 27, 44 (top), 45 (top), 46 (inset), 50 (inset), 52 (inset), 76 (inset), 88 (inset), 100 (inset), 102 (inset), 120 (inset), 130 (inset), 136 (t. left), 137
Thompson & Morgan 90 (inset), 124 (inset), 140 (t.right)
Unwins 19, 21 (left), 143 (left)
Mike Warren 15 (top & bottom), 21 (right) 23, 24 (bottom), 29 (top & right), 31, 35, 37 (top), 41 (right), 42 (bottom), 43 (right), 104, 136 (t.right), 142 (t.right)

PRINTED IN BELGIUM BY
proost
INTERNATIONAL BOOK PRODUCTION